FUNDABLE KNOWLEDGE
The Marketing of Defense Technology

Rhetoric, Knowledge, and Society

A Series of Volumes Edited by
Charles Bazerman

Winsor • *Writing Like an Engineer: A Rhetorical Education*

Van Nostrand • *Fundable Knowledge: The Marketing of Defense Technology*

FUNDABLE KNOWLEDGE
The Marketing of Defense Technology

A. D. Van Nostrand
Georgia Institute of Technology

LEA LAWRENCE ERLBAUM ASSOCIATES, PUBLISHERS
1997 Mahwah, New Jersey

Lawrence Erlbaum Associates, Inc., Publishers
10 Industrial Avenue
Mahwah, New Jersey 07430

Cover design by Kathryn Houghtaling

Library of Congress Cataloging-in-Publication Data

Van Nostrand, A. D.
Fundable knowledge : the marketing of defense technology / A.D. Van Nostrand.
p. cm.
Includes bibliographical references and index.
ISBN 0-8058-2122-8 (cloth). — ISBN 0-8058-2123-6 (pbk.)
1. Military research—United States—Finance. 2. Federal aid to research—United States. 3. Research institutes—United States—Finance. I. Title.
U393.5.V26 1997
623.3'068'8—dc20 96-33342
CIP

Books published by Lawrence Erlbaum Associates are printed on acid-free paper, and their bindings are chosen for strength and durability.

Printed in the United States of America
10 9 8 7 6 5 4 3 2 1

For Joan,
whose passion for ideas liberates mine.

Contents

Editor's Introduction

Charles Bazerman, Series Editor
University of California, Santa Barbara

In *Fundable Knowledge: The Marketing of Defense Technology*, A. D. Van Nostrand takes us on a remarkable journey into one of the largest knowledge production systems in history, one that has funded much of the academic and industrial research in the United States over the last half-century. Within the closed world of defense research, directed by government mandates for free competitive bidding, an unusual market has developed. Only those closely involved with this often-secret enterprise have had a detailed sense of how this knowledge market worked.

The market sustains itself on paper: legislation, budget resolutions, guidelines and regulations, announcements of initiatives, white papers, requests for proposals, contracts, reports, capability statements. Documents in these well-ordered genres articulate with one another in systematic regularity, carrying out the business of the negotiation, contracting, and production of knowledge within the constraints of law. Yet they also have provided the flexibility to innovate new technologies, as all parties seek after their notions of the possible and the advantageous.

The systematic interaction of these genres defines what knowledge is, what problems get posed, and what kind of knowledge gets produced. The process converts information and potential information into problem-relevant knowledge; moreover, as Van Nostrand points out, the process results in the competance of providers and purchasers to identify and carry out new knowledge ventures. Increasing competence is as much part of the knowledge produced by the system as are the specific pieces of information contained in final project reports. The capability statement displays most directly the production of competence. Van Nostrand, among his many accomplishments, draws our attention to the ubiquitous but almost invisible genre of capability statement, important in all the professions.

While the defense knowledge market has supported much of the economy by directing defense production and by providing new ideas for the civilian market, it has been kept alive only by the political desire for national security in military terms. In the post-cold-war world, however, national security is being reinterpreted in economic terms. In order to serve the civilian commercial market, this knowledge production system—built on a close communicative relationship between a small number of vendors in symbiotic relation with a single client, articulated through many agencies and subsections—must find a new way of doing business. Its entire system of genres by which knowledge production is conceived, contracted, and completed must be opened up to the needs and dynamics to which it was not originally designed to be responsive. The current stage of defense conversion, making existing defense knowledge available to commercial users, is only the first step in a much larger process of reordering the communicative patterns by which knowledge is produced.

Defense spending (often to our chagrin and embarrassment) has been one of the great machines driving and feeding funds to the development of science and technology, not just in the last half-century, but throughout history. Can we find a peaceful way to maintain serious research on the same scale without a military motive, driven only by social priorities of prosperity, amity, and social well-being, and by environmental health? Will research serve only the most immediate demands of corporations, aimed at products that can turn a profit in a few years, or can we also fund research inspired by possibilities of the future? The answers to these bold questions are to be found, if we are to follow Van Nostrand's lead, not in bold ideological statements, but in the rhetorical details of the systems by which knowledge is produced. The issue then becomes: Can we develop a highly articulated knowledge market that will produce the kind of knowledge we would want to have and that will produce the kinds of competence that will make our world a better place?

Fundable Knowledge: The Marketing of Defense Technology provides powerful tools for thinking about the relationship of the three themes of this series: rhetoric, knowledge, and society.

Preface

The journey on which you are about to embark winds through the domains of a vast knowledge factory. Its route reveals the ways in which these domains of social activity are related. They compound one another with enormous consequence. Successive discoveries along the route disclose a knowledge market that permeates the national economy, a market with both immediate and future implications. Yet with all its disclosures, the itinerary of your journey is more orderly than the one on which I embarked.

The route I traveled was digressive and often circuitous. It began in a course I was teaching at Georgia Tech and reached across campus to a dozen laboratories. It then led to myriad offices at federal agencies in Washington, DC, to meetings of professional societies in disciplines that deal with technology, and to the congressional office buildings. During the first few years of my explorations, new discoveries simply expanded the area of my ignorance. Each new log that I threw on the fire, or that fell into the fire by chance, increased the circle of darkness to be penetrated. By degrees, during the next several years, these discoveries began to reveal the patterns that *Fundable Knowledge* describes.

I could not possibly have arrived at these discoveries without the guidance of many caring persons who taught me along the way. I hope to thank them now by recalling the trail that they helped me blaze. And, with luck, this trail will also introduce you to the scope and sequence of this volume.

The trail began 9 years ago with a course in technical communication at Georgia Tech. A few of us designed the course based on Bruno Latour's proposition that science-in-the-making is a different order of reality from science viewed retrospectively. Intended for undergraduate engineering students, it focused on current research in science and engineering on the Georgia Tech campus; it took advantage of a learning resource distinct from the teaching laboratories. Our resource was the Georgia Tech Research

Institute (GTRI), a congeries of laboratories specializing mostly in related fields of electronics.

What we fashioned was an exercise in cultural anthropology. Welcomed by a group of project directors at GTRI, students in English 3883 became priveleged observers of ongoing laboratory activities. Their responsibilities were to assess what they observed and submit timely reports of their assessments. As their instructor, I had the same responsibilities, but my final report has taken considerably longer to complete. My teachers were my colleagues in this venture at Georgia Tech and at GTRI, notably Ron Bohlander, Devon Crowe, Thomas Gaylord, John Gilmore, Kenneth Knoespel, and Jack Lackey. Their insights into institutional research and development (R&D) and its implications have informed this study. As one of these mentors observed, the students in English 3883 were "spreading out across campus to study the natives."

The natives taught them startling things about the production of knowledge that occurs through the processes of R&D. They taught them first and foremost that R&D is a business, that it is funded and fundable. Without sponsorship, most R&D would not happen. At GTRI about 75% of the R&D is sponsored by federal agencies for the national defense. As customers, the mission agencies in the defense establishment buy their knowledge primarily from R&D vendors. Observing this laboratory environment, the students soon became aware of the distinctive presence of one federal defense agency or another. That presence is as continuous as the dial tone in a telephone. And it is the premise of *Fundable Knowledge*.

In addition to the preeminence of sponsorship, the students discovered another startling fact. They perceived a descrepancy between the announced purpose of an R&D project and what they observed happening in its implementation. They read the proposals to sponsors, written by the project directors, that had led to the contract awards for R&D. But what they saw the researchers pondering were not necessarily the questions that the proposals had addressed and promised to resolve. The researchers were engaged in smaller, unresolved questions occasioned by equipment failure, unexpected test results, or serendipitous discoveries along the way. These questions tended to be instrumental or contingent; they were ways of factoring the larger unresolved conceptual matters. Precisely as Latour has observed, R&D-in-the-making is untidy, not at all like R&D in the textbooks. That sense of process infuses *Fundable Knowledge*.

A third realization that the students reported (with patient help from the project directors) was subtler and probably more profound. It concerned the architecture of these questions. Whether conceptual, instrumental, or contingent, the questions were typically formulated as *problems*. Values that were unknown but sought-after were consistently being defined in terms of *goals* and *obstacles*, thereby perpetuating a focus on what yet needed to be done. The strategy of such problem formulation was to enable agendas for converting unknowns to a status of being knowable. Once students recognized this mindset, they began to see its evidence in all of the R&D projects they were observing.

This reductive habit of rendering unknown values knowable excited my curiosity. It is what germinated this book about the larger system of knowledge production of which GTRI is a part. It is what set me searching for some conceptual framework for understanding the systemic nature of knowledge production in the whole enterprise of defense technology. Early on, Charles Bazerman helped me wrestle with conceptual frameworks, and James Wiltse helped me comprehend the scope of the enterprise.

Developed in response to the Cold War, the enterprise has vast proportions. During the past 50 years, more than 60% of all U.S. government expenditure for R&D has been invested in defense technologies. The defense R&D community is composed of hundreds of thousands of engineers, scientists, and information specialists. Knowledge is their basic output. The knowledge they produce enables the development of weapon systems, and most of it is available for transfer to civilian applications by state and local governments, educational institutions, and commercial firms. Knowledge is the basic commodity of the defense R&D community, and the knowledge production is systemic.

The market for defense technologies begins with the procurement of R&D projects. As customers, the mission agencies in the Department of Defense (DoD) buy their knowledge primarily from R&D vendors. Together, customer and vendor engage in the collaborative design of R&D projects. Through an elaborate system of iterations, they exchange their interpretations of shared information and thereby produce knowledge. Probably no one person understands all of the systemic complexities of this process. But in my long search for patterns to explain it, I was immeasurably helped by James Bolos, Charles Church, J. W. Dees, Fred Dyer, Michael Kelly, Edward Roberts, Bill Smith, Richard Truly, and Leo Young.

I am also grateful to a cadre of veterans of the defense R&D community; they are several dozen middle managers currently or formerly employed by the Department of Defense or by defense contractors, who have shared their knowledge with me. They have understood that my purpose is to analyze a large social system, not to expose it or denounce it or lobby for it. They have been generously willing to assist in this purpose, to help me "get it right," as one of them put it. Getting it right meant detailing their own jobs. They were not concerned about the landscape of the larger territory, but their local anecdotes were invaluable. I thank these persons for their help. They are the "old timers" that I refer to in the book.

The iterative process of knowledge production in the defense R&D community occurs within the larger context of political economy. A broad vision of technological capability for defense is spread among many policy makers with differing notions of national security and differing political objectives; it is what I have called a *distributed vision*. Within this context of conflicting special interests, the iterations exchanged by the R&D customers and vendors depend heavily on how the parties position themselves; these iterations reflect intensive marketing and market research. My exploration of special interests soon expanded to the arena of national technology policy and to the corporate cultures of the organizations that collaborate in making it. I needed to understand those organizational relationships, and for their precise observations about them I am indebted to Daryl Chubin, Susan Cozzens, Marcel LaFollette, John O'Brien, Philip Stone, and Arthur Squires. With their help, I was able to see how the various cultures of "official Washington" bear on the defense establishment.

In the production of defense technologies, problems of economics, politics, psychology, and marketing are often mutually contingent. They form a universe of surprising proportions, with insubstantial borders that keep dissolving and opening up new and different disciplinary questions. Exploring this universe was rather like digging a hole in the sand; the deeper one digs, the more the sides cave in, and the wider the excavation. But in the process I had reliable advisors. Richard Barke, Miriam Drake, Geoffrey Eicholz, Melvin Kranzberg, Gary Lehmann, and John Lundberg all gave me valuable insights from the perspectives of their own disciplines, as did Joan Pettigrew, Alan Porter, William Read, David Roessner, Richard Teach, Dean Temkin, and Elizabeth Wadsworth.

The patterns of knowledge production for defense that evolved during the long years of the Cold War are still evident, still operating. But since the end of that confrontation, the institutional resources for such knowledge production have been taxed by the need to make the national economy more competitive. Under the policy of Defense Conversion, this need extends to the federal laboratories engaged in defense R&D. They are now enjoined to engage in the transfer of their vast store of technologies to the civilian sectors of the economy.

Fundable Knowledge addresses the crisis in R&D management that this mandate has caused. The process of technology transfer lays a major burden on the defense R&D organizations. The new mission of the federal laboratories engaged in defense R&D commits them to two different kinds of goals at the same time. These organizations must now transfer their stored technologies, and they must do so even as they continue to develop new technologies. Moreover, they must sustain both endeavors with limited, or even reduced, funding. Within these organizations the two endeavors are typically seen to be at odds. Deliberately disseminating technologies beyond their military applications contradicts the whole culture of knowledge acquisition that is embedded in the defense establishment.

But even as burdensome as it may seem, the concept of merely disseminating knowledge scarcely explains the delicate and complex operations that technology transfer entails. I am indebted to Clifford Lanham, Kathleen Hayes, and Randy Goldsmith for explaining the intricacies of this process. The commercialization of a technology originally intended for a different application is a transforming process, a process of producing new knowledge.

In this context, many knowledge workers assigned to the transferring of technologies are ill prepared for their new responsibilities. The transfer process entails a marketing mindset, which begins with a perception of the customer's needs. But as customers themselves for the past 50 years, the mission agencies have typically not nurtured this mindset. The resolution of this dilemma is the culminating idea of *Fundable Knowledge.* The marketing of defense technologies has taken a dramatic turn. A new social technology is evolving for the production of knowledge targeted to new uses; a new infrastructure is being grown to support innovation. A solution is inherent in the dilemma.

The strategy of formulating a problem that contains a solution is what set me off on this journey long ago. The strategy is manifest in the public documents of the defense R&D community. They are generic documents. Prepared by R&D customers and vendors for their mutual exchange, these documents record every formal stage in the shaping of defense technologies. They constitute a literature of needs and claims. Rhetorical analysis of their texts reveals how they actually enable the transactions that produce knowledge. I am indebted to Charles Bazerman, Carolyn Miller, Greg Myers, and John Swales for their perceptions of how such analysis can reveal the interplay of text and context. And I have tried to heed the chief canon of discourse analysis, which is to make the analysis transparent so that the discourse can emerge.

The only secret to writing that I know of is rewriting. And the best agent of that process is someone who will read the text and respond to it honestly. After years of engagement, I know what it takes to conduct such formative reading of a manuscript. The task is to find within it the clues of what it might become and then delineate the chasm between its grasp and is reach. The task is awesome, a supreme act of inference. Therefore, I am deeply grateful to the six readers of the penultimate manuscript of this book. They are Charles Bazerman, Michael Kelly, Marcel LaFollette, Clifford Lanham, Joan Pettigrew, and James Wiltse. From the perspectives of their different disciplines, each of them has committed to this enterprise that supreme act of inference along with a mother's patience. No author could hope for more.

As for the text itself, parts of *Fundable Knowledge* have appeared in other forms. Sections of chapters 6 and 8 derive from my essay, "A Genre Map of R&D Knowledge Production for the U.S. Department of Defense," in Aviva Freedman and Peter Medway (Eds.), *Genre and the New Rhetoric*, Taylor & Francis, 1994; they are reproduced here with the kind permission of Taylor & Francis Ltd. Sections of chapter 11 first appeared as an article entitled "Technology and Communication in Military R&D" in November 1993; they are used here with permission from *Technical Communication*, published by the Society for Technical Communication, Arlington, Virginia.

My text also includes material from other authors. I am indebted to Stephen D. Antolovich, David L. McDowell, A. L. Mullikin, and the Georgia Institute of Technology for permission to summarize selected text material of theirs that appears in chapter 9. I also thank Randy Goldsmith for permission to use his copyrighted matrix, "The Innovation Process: A Model for Commercialization," which appears in chapter 12. The matrix is an evolution of work by Mohawk Research, Inc., with the Department of Energy and the National Society of Professional Engineers' cooperative agreement with the National Institute of Standards and Technology.

Rereading this Preface once more before sending it to press, I am acutely aware of the company I kept during a long journey. And I am struck by how fortunate I have been.

—A. D. Van Nostrand
Georgia Institute of Technology

Glossary of Acronyms

ABM	Anti-ballistic missile
ACTD	Advanced concept technology demonstration
AFOSR	Air Force Office of Scientific Research
AEC	Atomic Energy Commission
AMC	Army Materiel Command
ARO	Army Research Office
ARPA	Advanced Research Projects Agency (see DARPA)
ATBMP	Army Technology Base Master Plan
ATP	Advanced Technology Program
BAA	Broad agency (or broad area) announcement
BMD	Ballistic missile defense
BMDO	Ballistic Missile Defense Organization
CBD	*Commerce Business Daily*
C^3	Command, control, and communications
C^3I	Command, control, communications, and intelligence
CICA	Competition in Contracting Act of 1984
CDI	Center for Defense Information
COTR	Contracting officer's technical representative
CRADA	Cooperative research and development agreement
CRS	Congressional Research Service
DARPA	Defense Advanced Research Projects Agency (see ARPA)
DCAA	Defense Contract Auditing Agency
DCTP	*Department of Defense Critical Technologies Plan*
DDR&E	Director, Defense Research and Engineering
DFARS	Defense Supplement, Federal Acquisition Regulation
DIOR	Directorate for Information Operations and Reports
DOC	Department of Commerce
DoD	Department of Defense
DTIC	Defense Technical Information Center

DOE	Department of Energy
FAR	Federal Acquisition Regulation
FFRDC	Federally Funded Research and Development Center
FLC	Federal Laboratory Consortium
FTTA	Federal Technology Transfer Act of 1986
FY	Fiscal year
GAO	U.S. General Accounting Office
GOCO	Government-owned, contractor-operated
GOGO	Government-owned, government-operated
GTRI	Georgia Tech Research Institute
HSC	Human Systems Center, U.S. Air Force
HDL	Harry Diamond Laboratory, U.S. Army
IAC	Information Analysis Center
ICBM	Intercontinental ballistic missile
IQC	Indefinite quantity contract
IRI	Industrial Research Institute
MAD	Mutual assured destruction
MILSPECS	Military specifications
MIT	Massachusetts Institute of Technology
MOU	Memorandum of understanding
NASA	National Aeronautics and Space Administration
NEL	Naval Electronics Laboratory
NDRC	National Defense Research Committee
NIST	National Institute of Standards and Technology
NSB	National Science Board
NSF	National Science Foundation
NRL	Naval Research Laboratory
NTTC	National Technology Transfer Center
NTIS	National Technology Information Service
NUWC	Naval Undersea Warfare Center
NWC	Naval Weapons Center
OMB	Office of Management and Budget
ONR	Office of Naval Research
ORTA	Office of Research and Technology Application
OSD	Office of the Secretary of Defense
OSRD	Office of Scientific Research and Development
OTA	Office of Technology Assessment
PE	Program element
PM	Program manager
PR	Procurement request

R&D	Research and development
RD&E	Research, development and engineering
RDT&E	Research, Development, Testing, and Evaluation
RFP	Request for proposals
RXD	Research and exploratory-development event
S&T	[Defense] science and technology
SCEL	Signal Corps Engineering Laboratory
SDI	Strategic Defense Initiative
SDIO	Strategic Defense Initiative Organization
SLC	System life cycle
SOW	Statement of work in an RFP
SSA	Sources sought announcement
SSR	Sources sought response
T&E	Testing and evaluation
TRP	Technology Reinvestment Project
URI	University Research Initiative
WRDC	Wright Research and Development Center

1

Introduction: Mapping the Territory

On June 4, 1944, near the harbor at Torquay, England, a large shed stood empty. Its former occupants, part of a U.S. Army headquarters company, had boarded one of the troop ships in the harbor, bound for Normandy. Inside the open shed door, propped on an easel, a chalkboard framed the orders of the day. It was about noon when I read them. Three precise orders were inscribed in chalked letters, which nobody had bothered to erase. The words were simple, but the message was metaphysical.

REVEILLE WILL BE 0330.
DRESS WILL BE BATTLE.
MORALE WILL BE HIGH.

I stood staring at this nonsequitur. Morale will be high? Where did that come from? What a remarkable piece of syntax. By themselves, the sentences were innocent, but together they made a quantum leap in meaning. Repeating one verb twice in the future tense had instantly transformed a procedure into a spiritual forecast. That construct totally inverted a well-ordered world. The inversion was the sort of trick a poet plays to plant a vision where it is least expected. Then a larger wonder occurred to me. Who decided that morale will be high? Whose vision was this, anyway?

I was in a hurry. After the sudden delay in the invasion plans, I had been sent ashore to pick up some winch handles for the minesweeping gear on our ship. My errand was urgent, yet I stopped and stared again at the message. Who could have transformed a procedure into a commitment so deftly? The quartermaster? The company commander? The military system, or the society that sponsored it? Whose vision was it?

Years later, an answer to that question began to work itself out in the study you are about to read. Fundable knowledge is generated by a large social system that functions by impersonal directives not unlike the orders of the day on that chalkboard. The system produces knowledge for national defense. Goals are set; they are conveyed by impersonal directives, and most of the people who work in the system do not know who wrote them. But goals are set and then issued in these institutional orders of the day. At the rate of a billion or more words daily, the directives are issued from thousands of offices, and they articulate bits and parts of a vision. Most of the millions of people who produce knowledge for the system read only a fraction of these directives. Yet everyone understands that a vision of some sort is being distributed among these millions of people, which they are then presumed to assimilate and to implement.

Fundable knowledge is primarily technological knowledge. It is a commodity. Within this socioeconomic system, it is bought and sold in the form of research and development (R&D). And somewhat like other commodities in the futures markets for oil or grains, for example, it is frequently bought and sold before it is even produced. This is partly why it is called fundable; it has future value. But unlike other commodities in futures markets, defense R&D knowledge is produced by buyers as well as sellers. In fact, it is produced by buyers and sellers together, not in collusion but in collaboration. As peculiar as it might sound, the buyers characteristically contribute to the knowledge that they purchase, and the sellers often give away some of the knowledge they produce. Buyers and sellers exchange knowledge in an economic market that is like no other in the world.

THE MAGNITUDE OF THE MARKET

The defense R&D market is large. In fiscal year (FY) 1995, the Department of Defense (DoD) obligated some $39 billion for R&D. A few billion dollars of this sum were transferred to the nuclear laboratories of the Department of Energy (DOE) for technology related to nuclear weapons. But even excluding that fractional amount, the DoD's budget for R&D was the largest single category of federal R&D spending. By 1995, there was already a history of this proportionate expenditure; it was highlighted in the chapter "Research and Development: Financial Resources and Institutional linkages," reported by Science and Engineering Indicators (National Science Board, 1993). Since 1981, according to this reference, DoD had provided more R&D funds annually (for both in-house and external research) than all other federal agencies combined.

This lion's share of federally funded R&D dates back to the early years of the Cold War. A special report by Ronald Meeks (1994) for the National Science Foundation (NSF) summarizes 40 years of federal R&D expenditures

TABLE 1.1
Federally Funded R&D for National Defense
and Civilian Functions: 1955–1994 (Five-Year Averages)

Years	*Defense*	*Civilian*
1955–1959	84.1%	15.9%
1960–1964	69.1%	30.9%
1965–1969	51.3%	48.7%
1970–1974	52.3%	47.7%
1975–1979	50.0%	50.0%
1980–1984	59.3%	40.7%
1985–1989	67.7%	32.3%
1990–1994	59.1%	40.9%
Average	61.6%	38.4%

from 1955 to 1995. The report compares annual R&D spending for defense and civilian purposes during those 40 years, and displays them in terms of relative percentages. Based on these compilations, an abbreviated set of averages reveals a general picture of the proportionate expenditures. From 1955 to 1995, defense R&D funding averaged 61.6% of the total federal funding for R&D.

A graphic representation of these averages reveals how the ratio has varied over the 40-year period, reflecting the national response to military threats. Defense R&D funding stood as high as 84% of the total federal R&D expenditure in 1957, the year in which the Soviet Sputnik was launched. It shrank to about 50% during the defense budget drawdown shortly after the Vietnam War. Then it increased again to nearly 68% in 1987, reflecting partly the U.S. engagement in Star Wars. By 1995, defense R&D still accounted for about 55% of the total federal R&D funding (see Fig. 1.1).

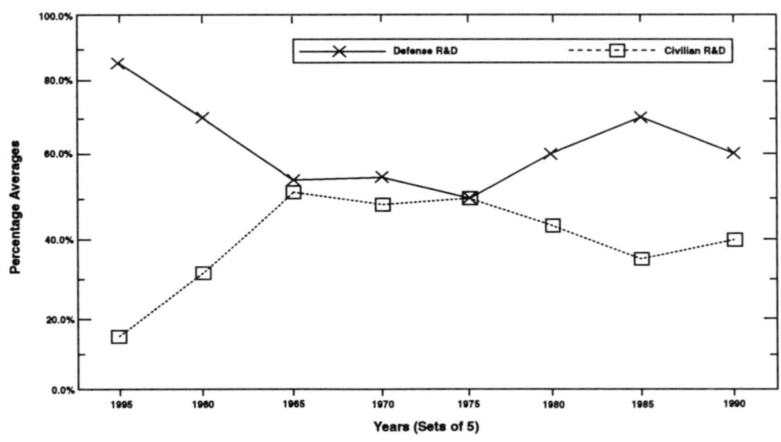

FIG. 1.1. Comparison of federal R&D funding for defense and civilian purposes.

A different perspective of defense R&D expenditure reveals not only its proportion of the whole but its magnitude as well (see Fig. 1.2). Defense R&D represented the greater share of a rising dollar commitment to federally sponsored R&D during the years of the Cold War. Reporting for the Congressional Research Service, Richard Rowberg (CRS, 1995) confirms the magnitude of the total commitment to defense and civilian R&D. He cites an increase, in current dollars, from slightly more than $1 billion in 1950 to about $69 billion in 1995. Policy initiatives that escalated civilian R&D costs included the Apollo space program during the 1960s and the energy crisis and the war on cancer in the 1970s. Then the defense buildup that began in the 1970s and continued for some 15 years accelerated the rate of increase in the total federal R&D costs. Rowberg's CRS report for Congress (p. 4) displays the relative dollar cost of these various escalating commitments.

The dollar volume of defense R&D and civilian R&D is a fair indicator of comparable magnitudes, but it does not reveal the relationship between the two domains. The present prospect is that they will become far more interactive than they have been during the past 50 years. Since the end of the Cold War, that prospect has become more significant to the national economy. The proposed reciprocity between the defense technology base and the national technology base is concurrent with the ongoing drawdown of the defense budget. Increased interaction between the two technology bases has evolved into a long-range program of defense conversion. As

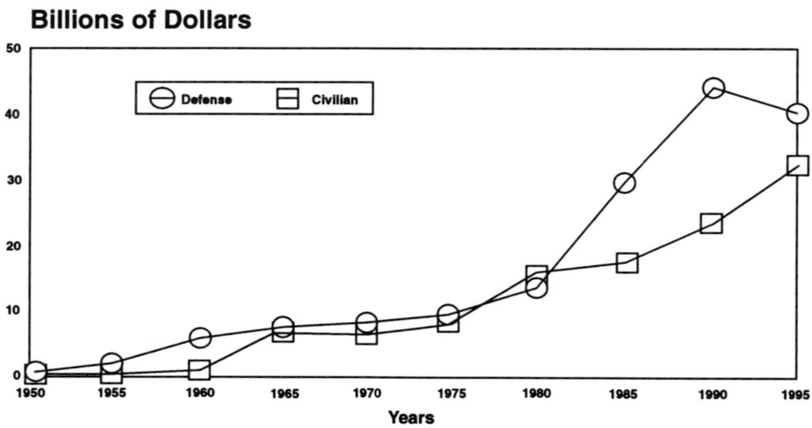

FIG 1.2. Federal civilian and defenses funding trends.

described in Adjusting to the Drawdown, prepared by DoD's Defense Conversion Commission (1992), the program represents a new vision of defense technology. In this vision, the acquisition of defense technology is seen as contributing to the competitiveness of the national economy.

As regards defense R&D, the official vision entails the merging of the defense technology base with the national technology base. In Defense Conversion: Redirecting R&D (1993), the congressional Office of Technology Assessment (OTA) offers a comprehensive description of the merger. It presents a broad range of initiatives as policy options available to Congress. By any measure, however, the merger is a radical undertaking; it entails abrupt departure from the conventional ways of doing business during the Cold War. Defense conversion and the envisioned merger of the two technology bases provide the focus of this study of fundable knowledge.

DEFENSE CONVERSION

Defense conversion after the Cold War represents a new national experience. It is unrehearsed. During World War II, the nation mobilized an entire economy for waging war, an effort coordinated, appropriately enough, by what was then the War Department. When peace arrived, a restive insecurity in response to Soviet military power provoked a period of sustained rearmament based on new technologies. A new system for acquiring those technologies evolved; it grew and became increasingly burdened with social and political constraints and Cold War biases. The military drawdowns that followed the hot wars in Korea and in Vietnam were only temporary pauses in this inertial acquisition system; although defense budgets were pared back after hostilities ceased, the defense acquisition process continued as it had before. The DoD went on acquiring knowledge in the same way.

The task now is to integrate DoD's specialized technology base, maintained for 50 years, with a national technology base intended for broader purposes. But the task is formidable for two reasons. The newly declared national technology base has yet to be configured, and DoD's technology base was developed in a culture that resists change.

Beginning in the mid-1980s, the vision of defense conversion was gradually articulated by a series of congressional acts and executive orders. In piecemeal fashion, new initiatives have been instituted. One is the mandate that DoD buy its technology on the open market wherever possible instead of having it custom made to military specifications (MILSPECS). Another initiative is the formation of joint R&D efforts involving the DoD with firms in the private sector for the joint development of knowledge products. The initiative includes a variety of working arrangements, especially formal Cooperative Research and Development Agreements (CRADAs).

The broadest initiative that affects the merger of defense and civilian technology resources has been the congressional legislation for technology transfer. The Stevenson-Wydler Act of 1980, then the Federal Technology Transfer Act (FTTA) of 1986, with its subsequent broadening amendments, have affected all federal laboratories. The legislation directs them all, more than 600 of them, to engage in transferring their knowledge products to local and state governments, to universities and other non-profit organizations, and to commercial firms. It is a bold venture. This version of technology transfer differs radically from the more general notion of transferring resources from the well-endowed societies to the underendowed. It is a highly refined concept of reciprocal exchange for mutual benefit, and it entails complex interactions.

The mandate for technology transfer challenges all federal laboratories to do business differently. About one in every eight of these laboratories is directly involved in defense R&D, and their common dilemma is acute. In the progressive reduction of defense budgets, military R&D organizations are fighting for their fair share of funding; some are fighting for their existence. Although technology transfer now has mission status, it conflicts with a venerable corporate culture. Within many of the DoD organizations, despite the mandate, the organized dissemination of knowledge is simply not seen as part of the mission; it is regarded as an inversion of priorities. For years these organizations have soaked up knowledge like so many sponges; now they must squeeze it out, even as they continue to soak up more.

Achieving this feat, requires a new technology of social means. With reference to the nature and the magnitude of these changes, this book explores the development of defense technology. It centers on the process of knowledge production; it examines kinds of knowledge and ways in which the defense R&D community produces them. The book parses the process of knowledge production, factoring it into component parts and observing their relationship to one another. And it addresses ways in which the process will need to be modified in order to achieve technology transfer, which is crucial to defense conversion. The rationale of the book lies in the relationships among all these elements of change.

Both the funding and the management of knowledge production engage the members of the defense R&D community in a great deal of explaining. Persuasion is their purpose, so rhetoric is critically important to the whole R&D enterprise. So also is the marketing of R&D. Since buying and selling knowledge entails continuous persuasion, both marketing and rhetoric are powerful forces in the development of defense technology. The domain of this study is a large social system, the defense R&D community, in which all of this interaction occurs. As in any large physical system, the interactivity in this large social system is complex, nonlinear, and full of variables. Through its continuous negotiations, text and context merge.

In sum, this book undertakes an analysis of patterns within a large social system whose knowledge output has national consequence. Its time frame is the 50-year period from 1945 through 1995, which covers the Cold War and the beginnings of defense conversion. In the search for patterns, I have attempted to disassemble the components of the system—rather like reverse engineering. The disassembly enables us to see how the parts fit together and function, and to learn thereby the significance of the massive change that is occurring. The patterns reveal how the business of defense R&D became institutionalized after World War II, what it evolved into, and how it produces knowledge. The patterns also provide a sense of how the defense R&D community is likely to cope with the dissemination of knowledge in face of the mandate for Defense Conversion. Since the patterns delineate a social system, this is a study of distributed vision, of a vision inherently subject to interpretation by those who implement it.

I recognize that my audience is likely to include both insiders and outsiders, both members of the defense R&D community and observers of it. So I am aware of needing to address some possible assumptions you might already have and to invite you to consider still others. If you are not a member of the defense R&D community, you might care to consider a new perspective of it. You might ponder the proposition that guns and bombs are derivative products; they are consequences of knowledge production, which is the basic activity of the defense R&D community. Another ponderable proposition concerns the nature of knowledge. Specifically, what we call knowledge is information that has been rhetorically organized. These two propositions about knowledge and knowledge production are premises that we will revisit often.

If, on the other hand, you are an insider, you already know the game; you have your own sense of the cliché about pushing papers in a bureaucracy. This particular bureaucracy is huge. Of the millions of people who are a part of it, only a tiny few have any sense of the system as a whole, and no one completely knows it. But whatever your assumptions about the system, here is a chance to take a fresh look at that part of it in which you are engaged. Chances are, you are either disseminating information or interpreting it—or both. So you are thereby engaged in the fundamental activity of the defense R&D community, which is the conversion of information into knowledge. Knowing how you contribute to the knowledge-producing process may give you a new perspective on the corporate vision you are helping to distribute.

For insiders and outsiders alike, this study has a common starting place in what we may call the universe of R&D. In the far-flung territories of this universe, scientists and engineers collaborate to develop new technologies. The universe is a social entity composed of organizations. One has to come to terms with this proposition at the outset; the universe of R&D contains organizations that interact with each other to produce knowledge.

COMING TO TERMS WITH DEFENSE R&D

Before the outbreak of World War II, the U.S. government sponsored some linkage of research and development in the private sector. Robert Schlaifer (1950) cites initiatives by both the Army and the Navy with respect to the design of aircraft engines during the 1930s. But the linkage of research and development was not formally part of federal technology policy until a few months before the bombing of Pearl Harbor. On June 27, 1940, President Roosevelt issued an executive order establishing the National Defense Research Committee (NDRC) of the Council of National Defense. A civilian committee of scientists and engineers under the direction of Vannevar Bush, the NDRC was an instrument for organizing a national research community that could be mobilized for warfare.

NDRC operated on a radically new premise for involving scientific research in the development of weapons. In dealing with the military services, it took the initiative in setting research priorities and assigning resources to support them. NDRC had enormous authority. In *Organizing Scientific Research for War*, Stewart (1948/1980) recalls that "NDRC asserted the right to exercise an independent judgment as to the projects which it should start as well as the method of attack upon them" (p. 18). This innovative protocol is explained in terms of the advances in science and technology that had placed new weapons development well beyond the normal expectations of military specialists.

But management innovation was not yet complete. Gaps soon occurred between fundamental research and the initiation of weapons procurement. An intermediate phase of engineering management was needed. So the NDRC was absorbed into a larger administrative entity, one that formalized the linkage of research and development. The function of the new entity was to provide an uninterrupted sequence of inquiry: from technological ideas, through engineering development, to the production of weapons. An executive order on June 28, 1941 mandated this new organization, the Office of Scientific Research and Development (OSRD), and the concept of a seamless R&D sequence was made official. In his foreward to Stewart's history, Vannevar Bush later alluded to the extraordinary innovation of this national, mission-oriented R&D system when he characterized World War II as "the first war in human history to be affected decisively by weapons unknown at the outbreak of hostilities" (p. ix).

In response to the Soviet military threat during the next 50 years, the R&D infrastructure was expanded and refined. Then, after the Cold War, some radical changes were initiated that affected the territory of R&D. Undertaken as part of the vision of defense conversion, they were confirmed in 1994 and incorporated into a new concept called the Defense Science and Technology (S&T) Program. This program exists conspicuously at the policy level, its purpose being to articulate the new priorities of defense conversion and to oversee their coordination at the operational level. The program has caused seismic shifts among certain jurisdictions in the management structure of DoD's acquisition and technology program.

Despite such change, the components of the R&D effort are generally much as they have been for the past 50 years. They constitute a complex network of organizations with missions that are more or less complementary. For organizational purposes within the military services, the R&D effort is extended to include the activities of testing and evaluation (T&E). So the whole succession of knowledge-producing activities prior to the production of technology has a long designation: Research, Development, Testing, and Evaluation (RDT&E). Although the letters in the acronym might tempt one to visualize these activities in serial fashion, a linear model would be inaccurate. For a better approximation, imagine a set of four ellipses, all nested together and all touching at the left-hand curve. Then label the inner ellipse "S&T," the next two larger ones "R&D" and "T&E," and the outer ellipse "RDT&E." Then visualize these four ellipses lying within a still larger ellipse that represents the all-encompassing acquisition system (see Fig. 1.3).

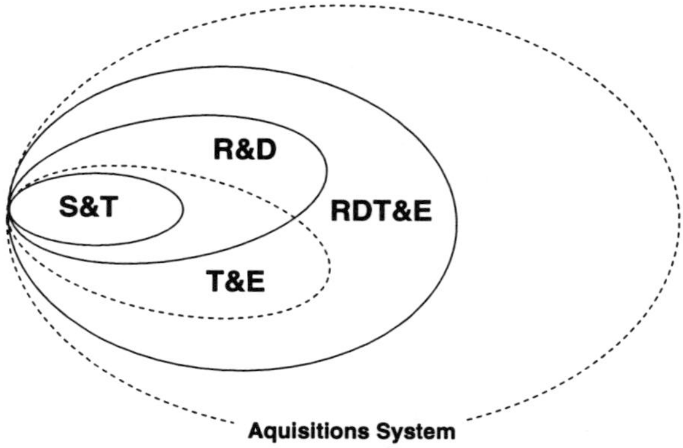

FIG 1.3. The defense domains of knowledge production.

Now for a closer look at that inner ellipse. The domain of S&T currently represents an annual public investment of about $8 billion a year, which is about 20% of the DoD's comprehensive R&D budget and about 12% of the federal government's total annual expenditure for R&D. (Despite changes in DoD's overall budgets through the years, the proportionate expenditures for R&D have remained relatively stable.) The domain of this inner ellipse contains the interactions that have the most leverage on the cost and performance of defense technology development. The buck *starts* here, you might say. It is the domain within which the most conspicuous patterns of knowledge production occur. Those patterns also occur in more distributed ways within the successively larger domains of R&D and RDT&E.

THE AGENDA AHEAD

Pattern analysis reveals how visions are negotiated during the production of knowledge. The analysis begins with a broad view. Chapter 1 provides a context for the evolution of defense technologies in the United States during the Cold War; it centers on the Strategic Defense Initiative, which overstates the prominent patterns of technology development that we will later encounter. As a caricature of the politics of defense R&D, the Star Wars episode provides a quick, magnified view of the way business is done in the enterprise of national security.

Following this contextual chapter, analysis proceeds by successive stages that focus on different sets of patterns. Chapters 2 and 3 address patterns in the knowledge content of technology and in the processes of knowledge production. Chapters 4 through 7 address the economic market for defense technology and the marketing activities which it entails. These are social patterns. Chapters 8 through 11 address the ways in which informal agreements become formalized; they track the public discourse of knowledge production in the defense R&D community and also the rhetorical patterns that characterize it. Finally, chapter 12 explores the contradiction of social patterns that defense conversion entails. Wherever appropriate, the chapters reach back into the Cold-War past to explain how certain patterns got started.

Each cluster of chapters highlights the development of defense technology from a different perspective. The rationale of their sequence is that each cluster reveals a set of patterns that make it easier to decode those patterns that follow in the ensuing clusters. The pattern analysis is cumulative.

A preface to each cluster of chapters presents an overview of the concepts germane to the perspective that those chapters address. So each cluster begins, as it were, by coming to terms with the next set of concepts—as we have just done with the concept of defense R&D. The first overview comes to terms with the nature of information, the nature of knowledge, and the process of knowledge production. The next two overviews highlight terms germane to two perspectives of defense technology that are related. One overview examines the components of the economic market for defense R&D; the other anticipates the rhetorical organization of information that the market generates. The final overview comes to terms with technology transfer, which entails a conflict within the culture of defense knowledge acquisition.

So much for sequence. It remains only to comment on two aspects of the evidence we will encounter. The first aspect concerns magnitudes that are expressed in dollar amounts. I have not found them particularly helpful in pattern searching. Fiscal record keeping in defense technology is subject to so many different reporting techniques that defining magnitude is scarcely an exact science. To begin with, there is no federal budget for R&D. As the Competitiveness Policy Council (1994) explains, "The R&D budget is highly fragmented in Congress, divided among nine appropriations subcommittees" (p. 20f).

In this fragmented situation, what we have instead of a budget is a total amount spent on R&D by all federal agencies in any previous year. These figures are later reported to the Office of Management and Budget (OMB) and duly published by NSF. I rely on that source and very few others.

The other aspect of evidence worth mentioning concerns its sources. Although the study contains no classified matter, I have occasionally encountered information that might be regarded as sensitive. As I mentioned in the Preface, the persons who have supplied it wish not to be quoted. Respecting their wish for confidentiality, I have not cited the notes that I made during my interviews with them. But I have found it helpful to make reference to some of the insights those interviews yielded. Although there is evidence aplenty in the public documents for pattern searching, it is often cumbersome. Invoking it would sometimes necessitate an inappropriate amount of time and space—and repetition—to cite the documents that contain it. On these occasions, to sharpen a point about which there is obvious consensus, I have quoted or paraphrased remarks of persons who wished not to be identified. Those persons are included among the several dozen veterans of the defense R&D community who have helped to educate me. I think of them as old timers. Hearing from them occasionally offers one other advantage. Their own spoken idiom conveys a sense of the culture that they represent.

As one old timer said to me when it was time to end our long interview, "Since we sat down, my people across the hall have been solving the system. I'd better go see what they've turned up." This is our signal to do the same.

2

Going Ballistic

"The history of the Cold War has been the history of the world since 1945." So begins Martin Walker's (1993) study of the conflict. "Communications and missile technology required the Cold War's battle zones to girdle the entire earth and its satellites to orbit above" (p. 1). The global conflict encompassed two hot wars, and it polarized most of the world's nations in support of one cause or the other. Excessiveness became commonplace. Maintaining a nuclear stasis entailed the most costly buildup of military technology in history; it bankrupted one economy and put a long lien on the future of the other. The buildup also yielded a vast knowledge base, having accelerated the rate of knowledge production far beyond what any peacetime economy could afford.

The last great confrontation of the Cold War, the U. S. Strategic Defense Initiative (SDI), was a hyperbole, an overstatement of the entire conflict. It magnified the interdependence of technology and national security, which was manifest in foreign policy, domestic politics, and the economy. The ambiguous vision of an ultimate shield against nuclear attack was broadly interpreted and reinterpreted, celebrated, denounced, and continually revised. The ultimate shield became a distributed vision, subject to many interpretations, which the component technologies of SDI both reflected and enabled. How defense technology happens, how it interacts with the various national expectations, as magnified by SDI, is the subject of this chapter.

The stalemate between the U.S. and the USSR had stretched over 38 years when President Reagan intensified it with his announcement of SDI. That announcement began the last technological confrontation of the Cold War. In a nationally televised speech on March 23, 1983, he proposed the development of a defensive shield in the sky to protect the United States from Soviet nuclear attack. He called on scientists and engineers to develop a system of counter weapons deployed on the ground and in space to seek out incoming missiles and destroy them in flight. Reagan's declared goal was to render nuclear ballistic missiles impotent and obsolete. Yet ten years later, almost to the month, SDI was mothballed. U.S. Secretary of Defense Les Aspin confirmed the "end of the Star Wars era," announcing a more moderate R&D program for ballistic missiles to succeed SDI. *Ballistic Missile Defense: Evolution and Current Issues,* a report of the U.S. General Accounting Office (GAO, 1993), provides a record of the SDI saga.

SDI became an instant rhetorical weapon. It combined two powerful ideas, a consummate national defense and the control of outer space. The force of it was evident in the vigorous reactions it engendered, not only in the Soviet Union and among the Western allies but within the United States as well. Conceptually monumental, it also had a spectacular political fallout. Rebecca Bjork (1992) refers to SDI as a symbolic containment of the nuclear threat. The authority of the symbol lay in its linkage of two abstract values—absolute defense and the control of space—with the authenticity of science and technology. Reagan's (1983b) invocation of this authenticity was explicit.

> Let us turn to the very strengths in technology that spawned our great industrial base and that have given us the quality of life we enjoy today. I call upon the scientific community in our country, those who gave us nuclear weapons . . . to give us the means of rendering nuclear weapons impotent and obsolete (p. 443).

These words became a celebrated endorsement of the efficacy of science and technology. Yet SDI was never a deployed weapon system. Limited by treaty agreement between the United States and the Soviet Union, SDI consisted exclusively of research and development. The implications of this obvious fact might not be as obvious as they deserve to be, for SDI was *assigned* to the status of R&D. In that status, the program did what R&D does, which is to manage the production of knowledge. Shortly after the President's announcement, DoD created a special agency, the Strategic Defense Initiative Organization (SDIO), to mobilize the endeavor of R&D in the thousands of technologies needed to develop a defensive space shield. As a result, SDI became a $30 billion crash program in knowledge production.

In retrospect, SDI can be seen as more than a spectacular technological episode. This remarkable program in knowledge production turns out to have been a surprisingly accurate index of the confrontations that characterized the Cold War. The military, political, and economic elements of SDI provide a set of indicators of the larger context in which it occurred. As the final episode in a long series of nuclear escalations, SDI yoked the ultimate technology to the notion of *absolute defense*. It became an exemplary statement of the commitment to definitive security that has driven American foreign policy. In this regard also, the presentation of SDI to the Cold-War world merged technology and rhetoric; R&D overflowed into politics.

In the aftermath of the Cold war, moreover, the SDI technologies have played a significant part in the complex process of defense conversion. They have left a legacy of knowledge to be transferred to the civilian sectors of the economy; the legacy entails building a broader and more integrated technology base than was possible in the Cold War. Not yet solved, the management of this integration is still being tested. In sum, all of these sword-and-plowshare issues have affected the sprawling nature of defense technology during the past 50 years.

A TECHNOLOGY FOR ABSOLUTE DEFENSE

SDI is a conspicuous indicator of the reciprocal escalations in weapon technologies that began with the atom bomb and occurred throughout the subsequent nuclear arms race. The escalations were rapid and precipitous. Within a few years of World War II, the Soviets had achieved nuclear parity and advanced beyond it. In a rapid series of events they exploded an atomic device in August 1949 and a thermonuclear device in 1951; by 1953 they had a deliverable thermonuclear bomb. With both sides in possession of the absolute power of destruction by 1956, the score in the arms race was then measured by the relative capability of delivering the bomb. In August 1957 the Soviets announced their successful test of the first intercontinental ballistic missile (ICBM), and two months later they launched Sputnik.

With the Soviet satellite visibly orbiting overhead, the United States was abruptly committed to some form of nuclear containment, which six successive U.S. presidents struggled to articulate. Robert McNamara, the Secretary of Defense for two of those presidents, described the standoff between the two contending powers as a condition of mutual assured destruction (MAD). Both sides increased their nuclear warfare capability; as a result, each side felt an acute sense of decreasing security. From this mutual condition a policy of bilateral deterrence evolved, confirmed by two major treaties between the U.S. and the USSR. The Limited Test Ban Treaty of 1963 concerned nuclear warheads, and the Anti-Ballistic Missile Treaty (ABM) of 1972 concerned a variety of ballistic missiles as well as multiple warheads. In effect, the two treaties confirmed the status quo of nuclear deterrence and the agreement to maintain it. This was the stasis in 1983 that prompted Reagan and his advisors to craft the Strategic Arms Initiative.

SDI was a technological overstatement, a pristine example of faith in the efficacy of sophisticated weapons. Harvey Sapolsky attributes such faith to the technological breakthroughs that occurred in World War II. In *The Polaris System Development* (1972), he cites the effect of those advances in technology on military thinking.

> [They] created the belief that the outcome of a future conflict would be determined solely by the technological power of the weapons that the adversaries could bring to bear in its first moments. Hence, the military had a strong desire to have always in hand weapons that were, technologically, the most sophisticated imaginable (p. 77).

Such weaponry had an obvious effect—Cold-War diplomacy centered on technology. Just as the Soviet Sputnik had shaped the confrontation twenty-six years earlier, SDI shaped the endgame. So it was with the nuclear events of the Cold War. Seen in retrospect, they formed an arc that resembled the trajectory of an ICBM; they began with a rapid escalation of offensive capabilities, entered a long, stable period of deterrence, and concluded with the rapid deescalation of the threat.

Technology and defense were the underlying motifs of the Cold War, and SDI combined them. Presented as a shield, as both a defensive and a technological system, SDI was a correlative of Cold War dynamics. It accentuated a foreign policy imperative that the United States had pursued for nearly 200 years: the concept of a total, unqualified defense against enemies—an absolute defense.

In the sense that a myth is a recurring theme that embodies some cultural ideal of a people, the American notion of an absolute national defense has been mythic. Its long history can be traced back to an Old Testament bias transplanted to the New World in the 17th century. Bjork (1992) reminds us of the circumstances in which the Pilgrims arrived on a new continent with their self-imposed isolation from Europe, and of their self-image of a chosen people with a God-given mission. Here was a durable mindset. Passed along to later generations, it could readily accommodate Thomas Jefferson's reservations about alliances with foreign powers and James Monroe's warning to possible intruders into the Western hemisphere. These public representations of self-sufficiency were compatible with the inherited sense of idealistic isolation.

James Chace and Caleb Carr (1988) see the preoccupation with absolute defense as the underlying theme in a long series of U.S. foreign policy decisions. They trace its roots to the British invasion of Washington D. C. and the burning of the capital city in 1814. The acute consciousness of total vulnerability that resulted from that humiliation began to generate a foreign policy that would extend the boundaries of defense farther and farther out. The Monroe Doctrine can be explained in terms of this strategy. So can the U.S. military occupation of the American West, the incursions into California, and the declaration of war on Mexico in 1846. The Spanish-American War in 1898 can be similarly interpreted. Such territorial expansions served the elusive goal of an absolute national security. Within this context, the control of outer space envisioned in the Strategic Defense Initiative can be readily understood.

But the notion of absolute security was ill defined, so it remained elusive. Territorial expansion for the purpose of defense served another goal as well. In the Spanish-American War particularly, it served economic opportunity. The American acquisition of the Philippines, as well as Guam, Cuba, and Puerto Rico opened up new markets. Yet territorial expansion ironically made the goal of achieving a viable defense more problematic. Walter Lippmann (1943) argued this point shortly before the Cold War began. He found the narrow pursuit of defense through expansion to be contradictory and self-deceiving, since it entailed commitments that could not be supported. A foreign policy is viable, he observed, only when power is brought into balance with foreign commitments. By a "foreign commitment" he meant an obligation outside the continental limits of the United States that might necessarily lead to war; by "power" he meant a military force sufficient to prevent a war or to win it. Maintaining a balance between power and commitment in foreign relations seemed as obvious to Lippmann as maintaining the balance between revenue and expenditure in constructing a sound fiscal policy.

In Lippmann's view, American foreign policy had become insolvent after the war with Spain. Reassured that our continent was a virtual island, we were comfortably committed to the total defense of the territory within its perimeter. But we lost sight of the fact that our foreign commitments had extended that perimeter thousands of miles in all directions; the area we had obliged ourselves to defend covered two thirds of the globe. We had enlarged the perimeter without providing the power to defend ourselves within it. Lippmann argued that the result of such an insolvent foreign policy was to render the nation unprepared to wage war or to make peace. America's floundering over the League of Nations at the conclusion of World War I and the Japanese attack on Pearl Harbor were prime illustrations of his geography lesson.

For whatever reasons, this nation's concept of military defense has had an abiding ambiguity. Our idealistic foreign commitments and our residual isolationism appear to have been at odds. Our legacy of isolationism accentuated a posture of defense, while our commitment to an expansive economic and social system engaged most of the world's societies; the defense perimeter became global. From this perspective, SDI was a correlative of the Cold War. A colossal initiative in the name of defense, it seemed to promise a return to the simpler times when borders were definable. But the ambiguity of SDI as a defensive measure was conspicuous. The Soviets saw SDI as clearly an offensive threat. In his account of the Cold War, Walker (1993) summarized their situation. "If successful, SDI would nullify the Soviet arsenal and restore that effective monopoly of nuclear threat which the United States had enjoyed in the brief years after Hiroshima" (p. 274). SDI, the defensive supermeasure, put the Soviets' own defense in jeopardy. The response of the Soviet General Secretary, Yuri Andropov, was an immediate assertion that the Soviets would never be caught defenseless by such a threat.

Reagan later hinted that the SDI technologies would be shared with the Soviets after the system had been developed. But since his own advisors publicly spoke otherwise, his tentative gesture offered no solace to the Soviets. Nor did it soothe the Western allies. They were confused by the whole defensive posture of SDI. Unlike the Soviets, however, they were dismayed by what they saw to be a withdrawal of U.S. protection. If successful, SDI might result in a separate nuclear peace between the United States and the Soviet Union, leaving Europe as a more vulnerable target for the Soviets' conventional warfare. From very different points of view, therefore, the allies and the enemy perceived SDI to be a disingenuous contradiction.

Reagan's rhetoric was not only an index of the Cold War's polarizing consequences; it conspicuously contributed to them. The situation in which he announced the initiative and the symbolic way in which he framed it made his rhetoric a weapon in its own right. His immediate audience was the American people, and his immediate political purpose was to counter the nuclear freeze movement that had begun in 1980; within two years it had spread evangelically across the nation and become a powerful lobbying force in Congress. The proposed freeze would be unilateral, in the uncertain hope that the Soviets would participate. At best, it would confirm the nuclear stalemate and lead eventually to nuclear disarmament and a negotiated peace. But such an outcome would presumably have to recognize the moral legitimacy of Communism, which Reagan vehemently opposed. Two weeks before his SDI announcement, he addressed the National Association of Evangelicals (Reagan, 1983a) to dissuade its membership from enlisting in the nuclear freeze movement. Casting the Cold War in moral terms, he warned his audience of "the aggressive impulses of an evil Empire" (p. 364). This inflammatory reference set the scene for his announcement. Two weeks later Star Wars became an entity.

THE COMPLEXITIES OF R&D MANAGEMENT

Although SDI was only a virtual technology when it was announced, it was based on a well-established concept of Cold War weaponry, the concept of an integrated system. The idea of a weapon as an integrated system of components was a revolutionary change in military technology that occurred early in the Cold War. Prior to this change, weapons had been developed piecemeal. Before integrated weaponry, for example, the atom bomb that the United States dropped on Hiroshima was limited by the accommodations of the airplane that would carry it to its target. Since the B-29 bomber already existed, the bomb had to be designed to fit the dimensions of the plane's bomb-bay doors. Moreover, the plane itself had to be modified in order to make the bombing run. It was even necessary to devise a separate loading mechanism. Because the bomb was too large to fit under the B-29, a pit had to be dug and equipped with hydraulic lifts to raise the bomb into the plane (C.G. Hibbard, 1995). But integrated weaponry changed all that. When the first Polaris submarine put to sea 15 years later, the boat and its nuclear missiles had been designed as a single integrated system.

The word "system" in this context seems beguilingly simple. For ease of reference, a tank (or a plane or a ship or a missile) is commonly regarded to be an integrated system. But the artifact itself, the engineered structure, is actually a system of systems, or more precisely, myriad systems and subsystems all working together. So the artifact is an integrated system in the most intricate sense imaginable. Each one of its component systems incorporates dozens of different technologies, and each subsystem of every component does the same. In this respect, SDI represented a massive integration of technologies.

As Sapolsky (1972) describes the concept of the *integrated weapon system*, its implications for management become immediately obvious. What that concept entails in the way of organization is staggering. "The weapon, its means of delivery, and . . . supporting equipment, *each the product of a separate organization*, were conceived as integrated packages to be produced to meet an overall optimal design" (p. 77, italics added). Imagine a jigsaw puzzle, each piece separately designed by a different supplier without the knowledge of what the whole puzzle would look like. Then consider the R&D task for such a construct: the kinds of interactions it would require among the different suppliers and the complexities of managing those interactions. Coordinating the myriad transfers of information that must occur in the design of integrated weapon systems, therefore, requires a massive management task. R&D management reached a new level of complexity during the Cold War.

One evident symptom of the complexity of R&D management is the long cycle time that ensues in the evolution of a weapon system from concept to production. The normal span of time is usually assumed to be from 10 to 20 years, during which new generations of different technologies are applied to the evolving weapon system. For ballistic missile defense (BMD), the cycle time was far longer, and it included several initiatives that predated SDI. The initial idea of BMD had occurred during World War II in response to the German V-2 bombs. In the absence of further threat, after the war, it was set aside in favor of designing defenses against long-range bombers. But that further threat suddenly occurred with the launching of Sputnik atop a Soviet ICBM in 1957. At about that time, using components currently available, the U.S. Army tested its first integrated BMD system. It was actually an assembly of systems consisting of a few ground radars, a set of high-speed computers, and a Zeus missile as an interceptor.

In 1958, a year later, Sputnik fever resulted in the chartering of the Advanced Research Projects Agency (ARPA), which immediately undertook Project Defender for BMD research. Project Defender included the physics of a body reentering the atmosphere and the design of sensors that could discriminate among targets. ARPA also sponsored a long-range R&D program in lasers and particle beams for use in directed-energy devices. Testing and evaluation of a prototype Nike-Zeus BMD system began in 1962 on Kwagalein atoll in the Pacific. The tests led to a new R&D initiative called Nike-X, which incorporated into the Nike-Zeus system advanced technologies previously developed in Project Defender. Nike-X was an integrated system; it included both short-range and long-range missiles, phased-array radars, and a parallel data processing system to integrate these other components.

The overlapping U.S. initiatives that preceded SDI were increasingly involved with a related set of technologies concerning the storage of the

ICBMs and the issue of their survivability in the case of a Soviet initiative. In turn, these logistic technologies were affected by issues of cost and by political decisions about deployment. Should the antimissile defense protect urban industrial areas in the United States, or should it merely protect the sites of the missiles themselves? The Johnson administration opted for the former; then the Nixon administration opted for the latter, in accordance with constraints on deployment imposed by the ABM treaty. The treaty also limited ballistic missile defense to the realm of R&D, which continued apace. The R&D yielded actual improvements in missile guidance for the interceptors and some progress in space-based lasers and directed-energy weapons. This was the situation when Reagan announced SDI.

POLITICS AND TECHNOLOGY

Reagan's announcement underscored yet another feature of R&D management that was characteristic of the Cold War—its extension to national politics. The intention to develop the technologies for a multilayered, anti-ballistic missile defense did even more than intensify the level of R&D effort. It broadened the scope of that effort to the congressional arena and, through the press, to the public. Star Wars became an instant entity. But that entity embodied the ambiguities inherent in both national preoccupations, ultimate technology and absolute defense. As a comprehensive program, SDI was still only a virtual system of technologies, a set of informed expectations. But it had immediate social and political status. Reagan's promise of an intensive R&D effort created new hopes and new budget allocations. For the time being, his promise of a defensive shield in the sky raised a rhetorical umbrella over the nation.

The umbrella represented a technological task of unprecedented proportions. Many scientists publicly challenged SDI. Even if system components could be designed, the whole enterprise would require an effective battle management system that could be fully operative within minutes. But clearly the task could not be accomplished until it was first defined. Meanwhile, funding was dependent on assumptions and objectives that kept changing. Was SDI meant to shield the people or the missile sites? Should it be nuclear or non-nuclear? Would it be deployed? If so, what about the ABM treaty, which prohibited deployment? Donald Baucom's (1992) history of SDI reflects an official interpretation of these issues. The myriad opinions about the issues complicated the process of engineering design.

Robert Lawrence's bibliography of SDI (1987) represents the intense debate in three broad categories: technical background, the case for SDI, and the case against it. But in reality the categories were mutually dependent, and the variables kept changing. The evolving architecture of SDI came under continuous review and revision, and issues of feasibility quickly became political. Predictably, Congress was ambivalent. Most of its members favored the ABM treaty which prohibited the deployment of anti-ballistic missiles. But jobs were paramount.

During the first two years of SDI, 77% of the contract money was directed to the states and districts represented by members of the armed service committees in both houses. This demographic pattern is reported by Sanford Lakoff and Herbert York (1989), who also discussed the intense scrutiny of the defense budget by large congressional staffs. Predictably, SDI was supported but underfunded, thereby stretching out the flows of money to designated contractors.

SDI was an exercise in the politics of technology. As expectations of goals, costs, and hazards were refined through many iterations, the architecture of the system evolved. The initial plan featuring clusters of interceptors housed in space based garages was deemed too expensive and too vulnerable. It was replaced by an architecture of *Brilliant Pebbles*, featuring thousands of autonomous interceptors in orbit, each with its own sensors and tracking devices. The review-and-revision process incurred a high cost. At the outset it was assumed that SDI would be limited by only the state of the art in component technologies. But it soon became evident that funding levels were also a constraint. Changes in architecture entailed changes in budget appropriations. As Congress became more involved in the review process, the balancing of technological priorities caused changes in direction, budget cuts, lost effort, higher program risks, and cancelled contracts. Large investments in R&D were scrapped when optimistic plans became unaffordable. In *Strategic Defense Initiative Program*, the GAO (1991) provides a retrospective view of the intense effort.

The normal detours and delays inherent in the defense R&D acquisition process were exacerbated in SDI. As interactions multiplied between Congress and the R&D community, the program took on yet another characteristic of Cold War R&D: voluminous reporting. I mentioned earlier that the output of R&D is knowledge. Such knowledge is typically delivered in the form of reports, generic documents that arrange and assess information from some given perspective to serve some given purpose. SDI quickly became a prodigious source of technical reports, composed from many perspectives and serving many purposes.

The reports began with Reagan's announcement of SDI. He immediately called for more information: a feasibility study based on the status of missile defense technologies throughout DoD and a policy study of the function of SDI in the exercise of political options. Thereafter, SDI was the subject of continuous reporting. Within DoD, the Defense Acquisition Board, a dozen or more high-level executives in military planning, closely monitored the changing architecture of the program. They were aided by reports from engineering teams within SDIO, which also managed its own program study. The Defense Science Board, the Office of Management and Budget (OMB), the Office of Technology Assessment (OTA), Lincoln Laboratories, and invited subject specialists all reported on the program. Some of the reporting entities issued repeated updates; *Ballistic Missile Defense* (GAO, 1993), for instance, cites 27 different GAO reports on SDI during the 10-year lifespan of the program. Most of these reports were based on other reported material; in the aggregate, they summarize thousands of engineering reports on the design of the components and subcomponents of SDI.

The Star Wars episode ended nearly as abruptly as it began. Early in 1990, after political changes at home and the reduction of the Soviet nuclear threat, SDI was radically modified. Its original goal of absolute national defense was replaced by a more modest set of objectives: to protect military forces in specific theaters of operation and to deploy this more limited defensive system in stages. Congressional policy imprints on these objectives were confirmed in the Missile Defense Acts of 1991 and 1992. In 1993, SDIO was renamed the Ballistic Missile Defense Organization (BMDO), and its mission was downsized to meet military threats in local theaters of operation. Like the Cold War itself, the intensive interactions of politics and technology that SDI represented were now history.

In *Affording Defense* (1989), Jacques Gansler has analyzed the context of Cold War technology within which the SDI episode occurred. Even as SDI was deescalating, he updated Walter Lippmann's argument about the unwitting discrepancy between power and commitment in U.S. foreign policy. One of the acknowledged experts on U.S. defense acquisition policies, Gansler refocused the discrepancy more precisely on military technology, grounding it to what he called the technological imperative. *Affording Defense* points out the incongruous "gap between America's actual national security capability and the posture statements and assertions of American policymakers" (p. 2). The book notes that within the first six years of Reagan's presidency, the defense budget nearly doubled, cumulatively committing almost a trillion dollars to increase America's military power. Yet by the end of the Reagan presidency, that alleged military power was still elusive, and the defense establishment was protesting over the downsizing of its budget.

Gansler has laid most of the blame for this lack of credibility at the door of the U.S. military planners for their unequivocal faith in technology. Analyzing the costs of this faith in terms of both money and extra security, he finds the money excessive and the extra security dubious. Military R&D is driven by a "technological imperative," he observes; "because we can have it, we must have it." R&D has been "based on what the 'state-of-the-art' allows rather than on what would be 'acceptable.'" Given this conviction, he observes, high cost is inherent in the design of weapon systems. This initial cost, in turn, makes production so expensive that purchase power shrinks and fewer weapons are produced. The resolute pursuit of the most advanced technologies, moreover, stretches the R&D interval farther into the future, again resulting in fewer weapons at higher cost. Finally, the fascination with technology has placed an emphasis on full-scale development of consummate weapon systems, overlooking crucial long-term development of the defense technology base (pp. 217ff).

One obvious realm of decisions pertinent to long-term development of defense technologies concerns their use for civilian as well as military purposes. This policy matter concerns direct congressional funding. Gansler has recommended a number of policy changes that would favor long-term technology development, and they appear to have been heeded; at least, they are compatible with what DoD has since formulated for converting defense procurement to the constraints of a peacetime economy. As will later be evident, these policy changes have entailed the radical revision of long-standing procedures in the defense acquisition system.

The defense drawdown has hastened these changes. Already mentioned, the conversion of the Strategic Defense Initiative Organization to the Ballistic Missile Defense Organization was the first obvious symptom. Just as the SDIO correlated many aspects of the Cold War, so the BMDO correlates the conversion to a peacetime defense policy. As successor to the SDIO, it has continued to engage in a long-range component of defense conversion: the active transfer of technologies to civilian sectors of the economy for commercial development. To this enterprise SDI has left a significant legacy.

DEFENSE CONVERSION AND THE TRANSFER OF SDI TECHNOLOGIES

The concept of *defense conversion* involves DoD in reaching out to acquire knowledge from a larger portion of the national technology base than heretofore. The outreach includes three means of merging defense R&D

with civilian R&D. One is the direct military purchase of technology products that civilian suppliers have already developed for other markets. Another is partnering with commercial firms in developing technologies that serve both military and civilian purposes. The third, and broadest, is the transfer of Cold-War technologies from defense laboratories to civilian organizations. All three of these consolidating moves are intended to strengthen the national economy as well as the nation's defense capability. In different ways, all three combine the forces of a free market economy with those of the regulated market for defense R&D; their common purpose is to accelerate the flows of knowledge between the two kinds of markets. The SDIO and its successor, the BMDO, have participated in all three of the consolidating initiatives.

The first initiative, the direct purchase of shelf items from civilian suppliers, reflects an immediate necessity. The reason for the change lies partly in the kind of technologies that DoD needs and partly in the long cycle time required to develop them through its cumbersome acquisition process. DoD continues to need the information-processing technologies that make integrated weapon systems possible. Its pioneering R&D efforts in the fields of computing, electronics, and optics during the Cold War led the world in these technologies, and its venture capital in their development has largely funded domestic commercial suppliers. But those suppliers can now develop "next generation" products within a few years, and sometimes within a few months, as compared to the 10 to 20 years it takes DoD to develop them. Continuing to rely on the conventional acquisition system, therefore, would result in the deployment of outmoded technologies and the loss of military capability. *Acquisition Reform* (GAO, 1994) describes the evolution of this program and the recommendations to be implemented.

DoD's second initiative for integrating two markets is the joint financial investment with civilian firms in the mutual development of dual-use technologies. The official vision of this integration is a joint policy statement (National Economic Council, National Security Council, and Office of Science and Technology Policy, 1995). Its implementation is set forth in *Dual-Use Technology* (DoD, 1995).

The dual-use concept is broader than the direct purchase of commercial products. It taps into the commercial sector's continuous stream of updated technologies at earlier stages in their development, especially during the design process. It also incorporates the idea of dual production, which takes advantage of flexible manufacturing technologies. The manufacturer can interpolate short production runs of a device for a military customer into larger orders of a similar device for commercial customers. The immediate advantage of the dual-use concept for the military customer is the opportunity to update current weapon systems with state-of-the-art technologies; a more strategic advantage lies in broadening the base of capable commercial suppliers. The commercial vendor gains as well, acquiring a new customer and also the benefits of the knowledge produced through that customer's own R&D effort. Their joint investment in dual-use technologies is intended to achieve a win-win situation.

The third form of outreach initiative is part of a broader federal mandate to all U.S. government laboratories to transfer their technologies to civilian sectors of the economy. The SDIO pioneered in this regard. Although technology transfer is an old idea, the federal mandate was a new formulation of it. Congress first specified a limited version of such transfer in 1980 and then broadened its provisions with subsequent legislation during the next 12 years. Designated recipients of federally owned technologies now include state and local governments, educational institutions and other nonprofit entities, and commercial companies. As regards defense technology, this dissemination of knowledge has notable consequence. Technology transfer is a categorical means of integrating the DoD technology base, which is highly defined, with the national technology base, which is far more comprehensive.

In defense conversion, as in Cold-War confrontation, SDI became an overstatement. The management of SDIO far exceeded the management of other defense R&D organizations in its early, full-scale commitment to making new technologies broadly accessible.

These included information technologies for software in artificial intelligence and laser technologies for etching in microelectronic components. Space-surveillance technologies were relevant to many different types of sensing and signal measurement. And tracking and pointing technologies were made available for commercial aircraft guidance and ground traffic monitoring. Ballistic Missile Defense (GAO, 1993) catalogues some of these technology transfers. SDIO initiated an online database of more then 2,000 transferable ideas, and its successor, BMDO, has maintained and updated the file. Already applied to the development of marketable products, some of those ideas relate to the production of semiconductors, computer disks, image processing software, refrigerants, batteries, surgical gloves, and beverage cans.

SDIO's exemplary achievement in technology transfer began with the appointment of its director, General James Abrahamson, who had been serving with the National Aeronautics and Space Administration (NASA). Borrowing NASA's model for technology transfer, Abrahamson set up a Committee on Education and Civil Applications, a board of advisors charged with identifying commercially applicable ideas. The advisors were information specialists, venture capitalists, and business managers; the ideas sprang from an inventory of ballistic-missile technologies that had been accumulating since the 1950s. The advisors convened forums in which commercial firms could present their technological needs and receive serious critiques of their plans to accommodate them. A commercial contractor, Bolt Beranak and Newman, was retained to coordinate these technology-transfer activities. In the ensuing years of SDIO, congressional favor was one incentive for this achievement. As Congress became increasingly involved in the status of SDI, it also continued to update legislation for broader technology transfer. So SDIO could use its technology-transfer efforts to advantage in positioning itself with the lawmakers.

What makes SDIO's achievement so unusual is the cultural context of technology transfer throughout the defense establishment. Like the other initiatives in defense conversion, technology transfer has generally run afoul of the perspective that guides defense acquisition. An unrelenting mindset serves the mission concept. Each one of the defense RDT&E organizations is formed around some particular military mission, and the acquisition of certain kinds of knowledge is paramount to carrying out each mission. But dispensing what has been acquired through dedicated labor and high cost is not yet broadly seen as a viable effort. "This stuff is not part of our mission" has echoed in the offices of many RDT&E organizations. There is a culture at stake, and the culture has its own inertial force and direction. Although compliance with the mandate to transfer technology is the order of the day, fully implementing it could take many years if left to the mission agencies. As a new R&D organization, virtually without precedent, SDIO was less constrained by the acquisition culture.

As a legacy of the Cold War, however, that culture remains a force to be reckoned with in both the acquisition of R&D and the transfer of technologies. The culture is a response to an environment that is both highly systemized and highly politicized, and it is inertial. These are the forces that shape the production of knowledge in the defense R&D community: the politics, the organizational system, and the culture that responds to them—all legacies of the Cold War. Within this context, we turn to the systemic production of knowledge that prevails in this community. Our first appointment is to inspect the nature of the knowledge that it yields.

COMING TO TERMS I

The Strategic Defense Initiative (SDI) was a crash program in knowledge production. It was solely an R&D enterprise, and it yielded an abundance of new technologies. SDI could also be described as a cumulative process. A fragmentary vision of a weapon system was modified by information from many sources, and the modified vision resulted in an abundance of new technologies. In a simple black-box model of the process, the input values would be labelled as information, and the output would be labelled as knowledge. In this model, both the information and the knowledge would include technologies.

The model is simple enough, but the fact that the input and the output both include technologies gives it something of a circular character. In fact, the roundabout nature of the model is critical to the way it works; we will deal with that at length. But first we must deal with the elements of the process. The words themselves—technology, information, and knowledge—are subject to so many meanings, so many usages, that they can get in the way of what they denote if we are not careful. So we must come to terms with them.

Indispensable tools, words can also be obstacles. T. S. Eliot (1963) represents the difficulty: "I gotta use words when I talk to you," Sweeney says to his friends in Eliot's poem, "Sweeney Agonistes" (p. 123). They are sitting around a card table talking about mundane things while Sweeney tries to explain the meaning of life. He is trying to come to terms with life, copulation, and death. His friends listen idly but do not hear;

they keep interrupting with seamy references to what has been going on lately. Determined to explain himself, Sweeney starts all over again after each interruption. "I gotta use words when I talk to you." Another of Eliot's poems obliquely explains what Sweeney is up against. The persona in "East Coker" wearily ponders the immensity of "trying to learn to use words." But "every attempt is a wholly new start," each venture is a "raid on the inarticulate," each try is undertaken "with shabby equipment always deteriorating" (p. 188f).

Coming to terms is one way of raiding the inarticulate. To use Eliot's analogy, the words themselves are the deteriorating equipment. Used and overused, belonging to everyone who employs them differently, they lose precision. The words that get blunted the most, like the ones that preoccupy Sweeney, are the high-level abstractions. Readily recognizable, they stand for so many different experiences that they elude precision. From time to time in this study I will pause to come to terms with high level abstractions that take on special meanings. At the moment, three words that are of this order demand attention. They are *technology*, *information*, and *knowledge*. Since they are crucial to the process of knowledge production, I will pause to say what I understand them to mean and how I will use them.

TECHNOLOGY

What we are getting at is the nature of knowledge production, and the knowledge in question pertains to technology. Most dictionary definitions of technology offer scattered generalizations; for instance, technology is the application of science, especially to industrial or commercial objectives; or it is the body of methods and materials used to achieve those objectives; or it is the body of available knowledge of use in fashioning implements, practicing manual arts and skills, and extracting or collecting materials. The common element of these definitions is the notion of instrumentality.

For teaching purposes, Steven Kline (1986) has classified technology according to four related usages of the term, all of them implying some kind of sociotechnical interaction. The most common usage of the term denotes *artifacts*: aspirins, bombs, telephones, airplanes. The term also denotes some *system of producing artifacts*, some combination of resources and methods, such as an assembly line or a set of laboratory procedures. Related to this latter meaning, but in a more abstracted sense, technology connotes the *skills or techniques implicit in the production system*; a certain manufacturer, for example, is said to have the technology to produce a given artifact. In the most extended sense, often taken for granted, technology represents *a whole social system, or collection of systems, for using an artifact.* If the artifact is an airplane, for example, the technology of its use includes laws and regulations, airports and schedules, insurance, contracts, fuel supply, training, even frequent flyer bonuses.

The derivation of the word encourages broad usage. The root seems clear enough, derived from *tekhne*, the Greek word for art or skill. But the suffix, *logy*, is more elusive. Derived from the Greek *logos*, meaning *word*, it has all sorts of extended meanings; it can mean the expression of something, or, very differently, it can mean the science or theory or study of something. I sought out a friend who is a physicist and a research manager to help me refine the notion of instrumentality in the word. I was trying to center the word, to find a core; together we pondered the matter. I remember the incident because it was so abruptly clarifying. On the table in front of me I had a dictionary, and I was reading a definition aloud when my friend interrupted me.

"You want to know what technology is? Technology is knowing when to pour the molten steel into ingots. It's knowing when the stuff is the right color for pouring."

I pondered that. Technology is knowing. That's the *art* or *skill* part of it, I thought. "But what about the *logos* part?" I asked. "What about the *expression* of that art or skill? Or what about the science or theory of it?"

"Look here. You're in a steel mill. The time is a hundred years ago." My friend was impatient. "You're the foreman at the hearth. This glass of beer is a ladle of hot steel." He pushed his glass toward me. "Is it ready to pour, or should you cook it some more? The manual tells you what the temperature of the stuff should be before you can pour it; some scientist has already figured that out. But does the manual say what the temperature is right now in the pouring area? Does it tell you about the draft blowing across the hearth? Or the extra time it will take for your short-handed crew to empty this ladle? You calculate these variables and then decide when the color of the steel is right for pouring. That's technology."

"OK, it's about knowing all these things and putting them together," I said. "That's the *tekhne* part of it, the art or skill. Now what about the *logos*, the word?"

"The word? Hell, I don't know. The *word* part of it is expressing what you know. It's telling your crew when to pour."

The parable of the steel mill helped to center the idea of technology for me, especially its usage within and about the defense establishment. That usage especially conveys two of Kline's categories: the system or process of producing something and the technique or methodology of producing it.

Lewis Branscomb (1995) blends these two senses of the word when he states: "A technology" is the aggregation of capabilities, facilities, skills, knowledge, and organization required to successfully create a useful service or product" (p. 186). One way or another, the foreman in the steel mill represents all of these elements. So Branscomb's definition is a useful referent.

Definition is a point of departure. It opens up the possibility of explanation, the chance for dialogue. Branscomb uses his definition as a referent in discussing national technology policy. Another, compatible definition provides John Kenneth Galbraith (1967)with the opportunity to explain technology in an economic sense. He defines technology as the systematic application of scientific or other organized knowledge to practical tasks. Its most important consequence, Galbraith explains, is in factoring any task into subtasks, so that existing knowledge can be brought to bear in resolving them one by one. The more sophisticated the task, the more complex and time consuming is the matching of its parts to existing knowledge. This explanation closely fits the development of defense technologies, as will soon be evident.

In order to incorporate the notion of development, we need to perceive technology as a process. If we add Galbraith's explanation to Branscomb's definition, we arrive at a restatement that seems to fit the need:

> Technology is the systematic application of available resources to the task of creating a useful service or product. The resources include capabilities, facilities, skills, knowledge, and organization, and the application entails a systematic factoring of the task into subtasks and matching them with available knowledge.

The matching of knowledge to tasks has economic consequence. The knowledge available for solving all the subtasks entailed in the production of goods limits the quality or quantity of goods that can be produced from available resources. In a commercial market, which is what concerned Galbraith, those limits affect the price of the goods. Defense technology, however, evolves in a different market, and there is a somewhat different consequence of matching tasks to knowledge. In the R&D effort for any new technology, matching knowledge to tasks entails not only cost but also elapsed time, which is critical to defense objectives. Galbraith's explanation is helpful in another way. It accounts for the state of the art at any given time, which can be explained as the degree to which the matching of subtasks to available knowledge has resolved those tasks. Once the matching of task to knowledge has been solved, that matching becomes the state of the art; further research and development can now be built on it.

INFORMATION AND KNOWLEDGE

Galbraith's explanation hinges on the major abstraction that this study addresses, which is knowledge. In Branscomb's definition also, knowledge is an explicit resource; it is also inherent in all the other resources that Branscomb cites: capabilities, facilities, skills, and organization. Knowledge is so central to technology that some operational definition of the term is needed. But any definition must accommodate an immediate detour. One of the first things we encounter about the usage of the word *knowledge* is its frequent confusion with *information*. The confusion is not surprising, given the widespread recognition that information has become the predominant force in determining societal relations.

There is already an extensive literature that identifies the prevalence of information and its consequences for doing business, managing organizations, producing wealth, conducting foreign policy, and waging war. Daniel Bell (1976), in *The Coming of the Post-Industrial Society*, was instrumental in shaping a new awareness of the power of information. Other widely circulated books develop this awareness: *Megatrends* (John Naisbitt, 1982), *The Twilight of Sovereignty* (Walter Wriston, 1992), and *The Post-Capitalist Society* (Peter Drucker, 1993). All these studies of the information revolution share the premise that an accelerating development of technologies has made it possible to disseminate information instantly and everywhere.

But common usage in a broader literature about information, much of it journalistic, tends to blur information and knowledge. We encounter phrases such as *information economy*, *knowledge capital*, *the dissemination of knowledge and information*, *the knowledge society*, *the knowledge base* and *the information base*. Contemporary American dictionaries tend to reflect the obfuscation. Some of them list information outright as one of the synonyms of knowledge. In other cases, definitions of information begin with the phrase *knowledge of* or *knowledge derived from* such and such.

To understand the process of knowledge production, however, we need to observe some distinction between these two words. If we recognize how information and knowledge are different, we can understand the dynamic relationship between them. One fairly obvious distinction is that we have much more available information than knowledge. Information is generic. Like mass or energy, it exists in many forms, and knowledge is one of those forms. Knowledge is a species, some unique kind of information, or better, some unique rendering of it.

Knowledge is information to which a certain value has been added in order to serve some given purpose, some need to know. It seems reasonable to speculate that the value added is some form of interpretation. A library is full of information, for example, but the information is inert until some client with a need to know something or other happens by. What needs to be known might be precise, or it might be vague. It might even begin with a global query, "How do I know what I want to find out until I see it?" In any case, from all available information, that need to know sooner or later focuses on whatever information will satisfy it. In the process of searching among information sources, one's need to know usually changes; guided by the information that is yielded up, much of it irrelevant, the need to know becomes more precise, more recognizable.

Any search among information sources entails the interpreting of information, which is generally a mixing and matching process. The information user has an *a priori* schema, some unique need for more information. At a certain point in the search, some of the information that has been accessed matches the user's schema, which has also been evolving. The user's interpretation may be an assessment of that information or a reorganization of it. In either case, the act of interpreting information formulates some particular significance about it. And that formulation is knowledge.

Producing knowledge is essentially a process of interpreting information, thereby adding value to it. The process can occur at different levels of consciousness, and it can be perceived in different ways. Different disciplines provide distinctive ways of talking about it. An economist might refer to information as a commodity, and to knowledge as a commodity with value added; the conversion of information to knowledge lies in adding the value. The value added might be measured as the difference between a sack of wheat and a box of Wheaties. A linguist might talk about predication, about putting a verb with a set of noun phrases to give them motion and a vector. In this case, the verb is the value added; it converts the phrase "cloudy and cooler" to a weather forecast. A poet might concentrate on the actual process of conversion and construct a metaphor about a chrysalis becoming a butterfly. In their different ways of wrestling the two abstractions down to earth, the poet and the linguist and the economist are all talking about the same evolutionary process. Knowledge production is this process.

In contemporary warfare the conversion of information to knowledge has immediate meaning. Electronic sensors deliver different kinds of digital information about a target to an electronic information processing center that fuses them into knowledge about the target. The process doesn't end there; the knowledge is transmitted as new information to a weapon that fires on the target. And all this is intended to happen in what is called near-real time. But the conversion process does not end there, either. Other sensors relay information back to the data-processing center about the condition of the target after the impact of the weapon. This information is then converted into new knowledge that will serve the decision of whether or not to continue the attack. Converting information into knowledge and transmitting that knowledge to some tactical receiver is a serial process of adding value in the form of interpretation; the process is essential to what is known in the U.S. armed forces as command, control, communications, and intelligence (C^3I). Defense technology is information intensive, but what it produces over and over again is knowledge.

So much for these terms. I propose that the conversion of information to knowledge in some serial manner is what underlies knowledge production in defense technology. The two chapters that follow extend this proposition with reference to the acquisition of defense technologies. Chapter 2 deals with the nature of those technologies, particularly with the knowledge inherent in them. Specifically, it describes a canon of defense technologies and the significance of measuring the knowledge that has evolved with them. Chapter 3 addresses the implications of chapter 2; it explores the process of converting information to knowledge that inheres in the development of defense technologies. It also delineates a pattern in the process.

Together, the two chapters reveal the systemic nature of defense knowledge production, which we will go on to explore. Thereafter, as the need arises, we will pause to come to terms with a few more concepts, which each ensuing cluster of chapters will then develop.

3

The Knowledge in Defense Technology

As the Star Wars episode demonstrated, the idea of absolute security in the United States is linked tightly to a belief in technological superiority. The linkage is a metaphor. It is an association of ideas with things, such as we commonly see in advertising, and the association is powerful. The linkage of absolute defense with technological superiority has the same kind of persuasive power, for example, that the marketers of Absolut Vodka have long exploited in their graphic advertisements. Each picture that blends the distinctive Absolut bottle into a familiar scene on the American landscape is a metaphor at work. The emotive force of a metaphor, the linkage of object and abstraction, lies in the association it makes in the viewer's mind. In the case of the vodka, the metaphor links the actual bottle to one's own feelings about the landscape, so that the feelings infuse the bottle.

There is also a metaphor at work in the public linkage of technology with the notion of absolute security. But this latter metaphor is of a different rhetorical order. Absolute security, the abstract part of the metaphor, can easily be asserted because it is readily idealized; it is an abstraction. But the technology, the object part of the metaphor, entails reams of explanation before it can be understood by the people who have to pay for it. And no wonder.

Defense technology represents a vast order of magnitude. It includes myriad technologies, each one of which is an aggregation of knowledge, skills, capabilities, organization, and facilities. Defense technology is also a social construct represented by thousands of organizations that manifest various parts of these constituent technologies. Even articulating the enterprise is a strenuous task, one that continually engages DoD. It is a rhetorical task, one that frequently entails persuading audiences; in pursuit of its task, as we are about to see, DoD imbues the concept of technology with a palpable shape and a sense of force. Our shortcut through this rhetorical effort is based on DoD's explanations of three aspects of defense technology. We examine first the strategy of defense technology, then the critical technologies on which that strategy relies, and finally the knowledge inherent in those technologies.

A PROFILE OF DEFENSE TECHNOLOGY

No single explanation of defense technology will suffice, since vastly different levels of understanding are required of those persons charged with developing the component technologies and those charged with paying for them. So DoD explains its mission in technology as persuasively as possible to different strata of audiences. Since we will have many occasions in this and later chapters to cite DoD documents, a reasonable shortcut is in order to help categorize them.

The varying density of evidence among these explanations depends on the needs and expectations of the intended readers. One group of audiences includes the President and other interested parties in the Executive Branch, congressional committees and subcommittees, and the general public. These audiences all need explanations that provide some definable context for whatever is being conveyed. Such accounts might be called *first-tier* explanations; they are what insiders within the R&D community communicate to outsiders.

When insiders explain things to one another, as in any discipline, they assume a mutual understanding of the context of what they are explaining. DoD's directives to its own constituent organizations, therefore, are less contextual and more detailed; they represent a *second tier* of explanation. In turn, the constituent organizations engage in a *third tier* of explanation, describing in still more detail their objectives and their needs to defense contractors.

Numbering annually in the hundreds of thousands, defense documents pertaining to technology vary widely in what they explain. The notion of merely three tiers of documents oversimplifies the rhetorical endeavor, but it provides a quick way of highlighting the defense technology profile. Within tier 1, two current documents are designed for different audiences. The first, *Defense Science and Technology Strategy* (DDR&E, 1994a), was intended for the broadest audience, including interested taxpayers. A brief but comprehensive overview, this primer is approved for public release and unlimited distribution. It articulates technology as a strategic element of defense. The second document, the *Department of Defense Critical Technologies Plan* (DDR&E, 1991), presents a profile of defense technology; it defines specific technologies and their relationship to the defense function. This document also provides us with an opportunity to examine the knowledge base of the constituent defense technologies.

A third document, issued much earlier than the others, provides us with a chance to examine the quantification and measurement of such knowledge; specifically, it developed a precise means of determining increments of knowledge in the ongoing development of a technology. This earlier document is *Project HINDSIGHT* (1969), the final report of a study that the Director of Defense Research, Development, and Engineering (DDR&E) undertook nearly 30 years ago. Although this report is a tier-2 document, its findings were so dramatic that it was widely distributed beyond the defense community. The quantification and measurement of the knowledge inherent in technology, which *Project HINDSIGHT* yields, enables the analysis of the flows of knowledge.

DOCUMENTING TECHNOLOGY AS A DEFENSE STRATEGY

Defense Science and Technology Strategy (DDR&E, 1994a) is the current edition of a short primer that explains the linkages of Science and Technology (S&T) in the context of the national defense posture. Its basic theme, "options for decisive military capabilities based on superior technology," (p. 1) is stated by the Secretary of Defense, and thereafter the primer explains it. The rhetorical strategy lies in the logic of its sequence. First we are told of the needs, then the necessary capabilities for meeting those needs, and finally the problems to be solved in order to develop those capabilities. Like a tight legal brief, the primer conveys a sense of inevitability in the way each aspect of the argument leads to the next.

The argument begins with the need to maintain superior military capability, and it proceeds to a clarification of the crucial role of technology in maintaining that capability. The reasoning evolves in stages. The current defense posture is defined by two kinds of security, military and economic. The military threat is no longer as predictable as it was during the Cold War, so responding to it requires a flexible force structure. The need for economic security requires the defense establishment to support the economy wherever possible. The combined needs require the conversion to a smaller, more flexible force structure that can be effectively supported. To maintain military superiority with fewer resources, therefore, technological superiority will be needed as a force multiplier.

That is the rationale for defense technology. DoD's objectives for its use as a force multiplier point to the kinds of knowledge that will be needed. The more specific objectives are of two kinds: in nuclear warfare, to counter the threat of mass-destruction weapons; in conventional warfare, to move armed forces globally and to engage an enemy decisively with minimum casualties and damage—all at a reasonable cost. The DoD primer explains that these objectives entail the use of technologies for the maintenance of military advantage in the dimensions of both space and time. Specifically, the control of space will enable uninterrupted access to global information. Managing the flow of that information effectively can maintain near-real-time knowledge of the enemy; it can also enable near-real-time communication of that knowledge to all deployed forces. As stated, the technology objectives concentrate on the flow of information; as implied, information technology will convert information to knowledge in near-real time.

Given these objectives for technology development, large problems become evident; characteristically they are called challenges in the military idiom, and these particular challenges involve the management of DoD's Science and Technology (S&T) Program. At this point, a gloss of the generic phrase Science and Technology is appropriate. Variously used, the phrase denotes both a concept and an organizational entity. Conceptually it denotes the front end of the R&D process. In this sense, Science and Technology is a domain of three parts: (a) research in science and engineering, (b) exploratory development of concepts derived from research that might serve military technology, and (c) early stages in the advanced development of weapon systems that might use those concepts.

Organizationally, the Science and Technology Program, usually called the S&T Program, has responsibility for this domain of activities. Like any other program, it is a planned sequence of management and budget decisions. Initiatives in the S&T Program are undertaken by the three military departments and by a few agencies within DoD. Coordination of their efforts is delegated to DDR&E. These few organizations are the principal actors in the S&T domain, although they are supported by other federal R&D organizations beyond and within DoD. Within DoD they are webbed with the larger infrastructure of Research, Development, Testing, and Evaluation (RDT&E).

The strategic importance of the S&T Program, as Gansler (1989) has pointed out, is the enormous leverage of front-end R&D over the design, and consequently the cost, of subsequent weapon systems. Having carefully charted these relative costs in the life cycles of its own technological systems, DoD has confirmed the pattern of such leverage. *Defense Science and Technology Strategy* (DDR&E, 1994a) presents this leverage as a management challenge to reduce costs.

> Because 80 percent of the life cycle cost of a system is determined by the first 5 percent of the life cycle costs—i.e., during the concept and preliminary design phase—affordability must be a key technology and design objective (p. 25).

Not surprisingly, this particular challenge dominates the longest chapter of *Defense Science and Technology Strategy*. Addressing taxpayers and their congressional representatives, DoD is careful to explain its efforts to lower the cost of such development. Some of the efforts address the direct costs of technology development, costs pertaining to supply, manufacture, and maintenance. Cost-saving efforts include the purchase of component parts from commercial markets, the use of automatic simulation methods in system design, and the development of new technologies to maintain equipment. Other efforts address the indirect costs of managing technology development. These include downsizing and reorganizing the larger network of organizations—the broad infrastructure of RDT&E—that supports the S&T Program.

Included in the RDT&E infrastructure are the DoD laboratories; research, development, and engineering centers; test centers; university research centers; various federal laboratories; and Federally Funded Research and Development Centers (FFRDCs). These organizations vary widely in size, function, and geographical location. They also have their own management structures. Some of them are government owned and government operated (GOGOs). Others, although government owned, are managed by contractors, which might be universities or private sector firms; that is, they are government owned and contract-operated (GOCOs). Not only are these dozens of different elements of the RDT&E infrastructure managed differently. They have their own missions to fulfill and their own different organizational cultures as well.

More can be said later about this RDT&E infrastructure, but a word in passing is appropriate. RDT&E is an immensely complicated network of organizations. As I mentioned a moment ago, defense technology is partly a social construct; it is an amalgamation of constituent, knowledge-based technologies and knowledge-producing organizations. In the organizational domain of defense technology, the S&T Program is nested within the whole

RDT&E infrastructure. But the relationship between the part and the whole gets confusing, especially when one tries to track the distinction between *program* and *infrastructure* in the nomenclature of official explanations. *Defense Science and Technology Strategy* for example, declares that "The S&T infrastructure itself, with its ability to act and react, is an important product of the RDT&E program" (p. 35).

> (It is worth pausing here to parse this statement because it is so typical of corporate prose in the defense establishment. That the S&T "Program" is a "product" is an interesting confirmation of the fact that we are dealing with a social construct. But it is curious that the S&T Program is also called an "infrastructure," and that the RDT&E infrastructure is also called a "program." Such circular nomenclature is a fair indication that the vision from atop the management hierarchy sometimes gets blurred; it is not always clearly comprehended by those who have to explain it.)

Defense Science and Technology Strategy is an overview. It is accompanied by a longer, denser tier-2 document that need engage us only briefly. Entitled *Defense Technology Plan* (DDR&E, 1994b), the companion volume explains how the development of defense technology will be implemented. It describes management decisions pertaining to the investment of resources in the constituent technologies, the assignment of organizational responsibilities, and the projected stages of development. The document is divided into 19 technology areas. Each area is described in terms of its scope of activity, its relationship to other technology areas, the rationale for investment, and the R&D organizations that it will engage. Less for taxpayers and more for Congress, the *Plan* is a general forecast of budgetary needs.

DOCUMENTING THE KNOWLEDGE IN DEFENSE TECHNOLOGIES

The next document in this profile is the basic directory of constituent defense technologies, the *Department of Defense Critical Technologies Plan* (*DCTP*). Regularly updated and approved for unlimited distribution, it is issued by the National Technical Information Service (NTIS). It assesses the state of the art in technologies that are critical to both national defense and the national economy. Comprehensive in nature, the *DCTP* is the result of broad inter-agency planning. It is prepared by the Department of Defense in consultation with the Military Departments, industry associations, and other federal agencies engaged in R&D activities. Among those other agencies are the Department of Energy (DOE), the National Aeronautics and Space Administration (NASA), the National Institute of Standards and Technology (NIST), and the National Science Foundation (NSF).

More than a directory, *DCTP* is a blueprint for development. First issued in 1989, it has become a cumulative planning tool in the evolution of critical technologies. *DCTP* annually updates a 10-year projection of DoD's technology priorities, the S&T programs that support them, and the dollars to be allocated to the respective Program Elements (PEs) that are assigned to implement their development. Like the *Defense Technology Plan*, it guides the congressional allocation of funds by setting program priorities and forecasting the dollar costs needed to meet them. But it goes beyond that other document in both depth and comprehensiveness. Several hundred pages long, *DCTP* describes the technologies at length and specifies details of their development.

Most of its bulk is an annex that profiles each of the critical technologies. Each profile includes the technology's intended military payoff, its dual-use possibilities, and its relationship to other critical technologies. As a program forecasting tool, the document uses the milestone, a standard unit of measurement in long-range planning. For each technology there is a set of objectives and a list of milestones to mark the implementation of each objective.

Consider, for example, the development of an unmanned land vehicle. The first milestone might be the capability of automatically starting and stopping; the second milestone might be the capability of turning at certain fixed speeds. Ensuing milestones might entail changing speeds, sensing obstacles, and turning to avoid them, with each new capability being attached to some future calendar date. Each milestone is a designated achievement in technological capability, and it is attached to a given fiscal year in the overall 10-year projection. In this respect, *DCTP* is a clear example of the systematizing of knowledge production.

To all intents, *DCTP* is the canon of defense technologies. Given its 10-year span, the list itself is relatively stable. Except for the updating of milestones and the addition of one or two new technologies, the corpus has remained the same since its inception in 1989. Although the canon can be examined from the perspective of any one fiscal year, FY 1991 offers some added value in historical terms. In that year, federally funded R&D for national defense represented about 60% of total national R&D expenditures. But the numbers were changing. The percentage for defense R&D

expenditures had been steadily declining since 1986, as consistent with planned reductions in overall expenditures for defense. Yet in FY 1991 the *DCTP* forecasted a 5-year increase in projected funding for the defense critical technologies.

For that fiscal year, *The Department of Defense Critical Technologies Plan* (DDR&E, 1991) profiled 21 critical technologies, seven of them serving warfare exclusively. The 15 others, intended for broader military and civilian applications, are listed here. (The list is annotated in Appendix A.)

1. Semiconductor materials and microelectronic circuits
2. Software engineering
3. High performance computing
4. Machine intelligence and robotics
5. Simulation and modeling
6. Photonics
7. Passive sensors
8. Signal and image processing
9. Data fusion
10. Computational fluid dynamics
11. Air-breathing propulsion
12. Composite materials
13. Superconductivity
14. Biotechnology
15. Flexible manufacturing

These are conspicuously information technologies. All of them use information in one form or another, and more than two-thirds of them directly concern the flow of information: its storage and recall, its transmission and transfer, and its monitoring or control. Moreover, the canon emphasizes virtually the same developmental goals for each one: higher speed, lower cost, and greater reliability. Another common element is the high degree of interdependence among the technologies. An advance in any one of them depends in part on parallel advances in a variety of others. By nature they are all instrumental; they all fulfill some need or set of needs. But some of them especially contribute to the development of other technologies; they are what Branscomb (1995) refers to as infrastructural technologies.

The first two in the canon, electronic circuits and computer software, are obviously seminal in this respect. Their derivatives include simulation and modeling, robotics, and flexible manufacturing. And there are degrees of derivation from one technology to another. Signal and image processing, for example, depends on at least a half dozen other technologies in the canon. These include semiconductors and microelectronic circuits, computer architecture, robotics, simulation and modeling, passive sensors, data fusion, and possibly others. So the canon is more than a compilation. This core of interdependent technologies is a synergistic web of continuous exploratory undertakings.

These technologies confirm a history of military influence in systematizing knowledge production. A brief sketch of two of the technologies sufficiently demonstrates the pattern. One of them is the seminal technology of solid-state electronic circuits, which began during the 1930s. To accommodate the need for information transfer in a battlefield environment, the Army Signal Corps was committed to miniaturizing electronic circuits. Before the outbreak of World War II, the Signal Corps Engineering Laboratory (SCEL) had already patented a number of processes for the fabrication of a ceramic-based circuit. So when a solid-state amplifying device was invented at Bell Telephone Laboratories in 1947, the Signal Corps was ready to combine forces. Misa (1985) describes the ensuing joint enterprises of the Signal Corps and Bell Labs during the next 10 years. At the Signal Corps' urging, joint R&D efforts resulted in the development of sophisticated transistors. In its quest for a transistor capable of amplifying high-frequency signals, the Corps also marshaled the engineering resources of Western Electric, General Electric, Raytheon, and RCA.

Summarizing the means by which mission objectives were implemented, Misa points to the consequences for the entire domestic transistor industry. In addition to influencing the size and structure of the industry through subsidies for plant construction, the military services conclusively influenced the production of knowledge. By setting standards for transistors, they biased the industry toward the development of certain types of devices. They also sponsored applications studies, conferences, and publications to the trade, which insured the rapid dissemination of information (p. 255). These means of interaction, I might add, also speeded the interpretation of the information that was being disseminated.

The critical technology of data fusion reflects a comparable influence of military organizations in systematizing knowledge production. David Allison (1985) describes the circumstances. In 1955, when the Navy started to update its mechanized shipboard data handling, a new technology of digital data correlation rapidly evolved. In that year the Chief of Naval Operations undertook a study of the comprehensive shipboard requirements for what would become the Navy tactical data system. To implement the system, the Bureau of Ships set up a project office to coordinate prime contractors in developing the components that the Naval Electronics Laboratory (NEL) would then integrate. The original contractors included Remington Rand, Hughes Aircraft, Collins Radio, and Ford Instrument Company. Some measure of the subsequent development of data fusion under the Navy's guidance is confirmed by the canon of critical defense technologies. By 1991, as set forth in the *DCTP*, this technology had become the province of the Army and the Air Force as well.

The model of mission-oriented R&D activities correlating the capabilities of many private-sector sources has a long history. It represents a durable policy of defense technology management.

MEASURING THE VALUE OF KNOWLEDGE

The common element in the components of any technology is knowledge, be it scientific, engineering, architectural, political, or organizational. Knowledge is not only an explicit part of the aggregation of resources that Branscomb (1995) cites. It is also implicit in the capabilities and skills and facilities and organization that are part of any given technology. Louis Tornatzky and Mitchell Fleischer (1990) remind us "Without knowledge there is no technology." The knowledge base of the tool or tool system that we call technology, moreover, is hybrid; it is essentially a cognitive construct in a social setting. "All technologies consist of knowledge about cause-and-effect relationships embedded into the physical tool, combined with knowledge about how to use the tool, all of which is embedded in a social context" (p. 13ff).

The knowledge inherent in technology lies chiefly in the design process. This proposition is explained from an engineering perspective by Walter Vincenti (1990). Analyzing what he calls "the internal logic of technology," he focuses on the design process as a set of problem-solving activities. Problem solving entails iterations, trade-offs, instrumental research, and other detours en route to achieving a workable set of plans. But whatever the design problems might be, he says, their solution depends on knowledge. When knowledge is not at hand, it has to be generated. "Either way, the needs of design impel, condition, and constrain the growth and nature of that knowledge" (p. 206). Technology incorporates knowledge.

Knowledge is a dynamic value in technology development. It is enabling. Once achieved, it is accessible to new interpretations; it becomes the means of further discovery. In technology development, knowledge increases incrementally. For purposes of planning, and acquisition those increments must be assessed. And to be assessed they must be measured. So the task of measuring knowledge is significant, and it requires management tools of a special nature.

The *DCTP*, for example, cannot measure knowledge, notwithstanding its precision in other matters. As a management tool it is a forecasting device; it sets goals, projects budgets, and designates broad organizational responsibilities for developing constituent technologies. Serving all these purposes, it is a structure built out of milestones, projected increments of capability to be achieved by fixed deadlines.

Milestones are anticipated stages of performance, useful for planning and the allocation of investment. But milestones do not measure knowledge; projected estimates are the wrong tool for that. For measuring knowledge, some retrospective form of observation is needed.

In 1965, DDR&E designed such a protocol for measuring the input of new knowledge into the technology of a weapon system. The purpose of the protocol was to assess the cost effectiveness of the overall R&D program during an extended period of time. Incorporated into an exercise known as Project HINDSIGHT, this measurement protocol was a retrospective study, based on the idea that a weapon system evolves through successive generations. With the aid of participating consultants, ad hoc teams of military and civilian DoD personnel were organized to study successive generations of selected weapon systems and to analyze the increments of knowledge that had enabled each new generation to occur. The resulting final report, *Project HINDSIGHT* (DDR&E, 1969), had such widespread impact on the national R&D community that it was approved for public release and unlimited distribution.

Focusing on 20 selected weapon systems that had evolved between 1946 and 1963, teams of specialists referenced critical points at which each technology had been advanced. Each interval between these critical points represented an increment of knowledge that informed the next generation of that technology. More precisely, they identified what they called RXD Events, each of which represented the incorporation of a novel idea into a particular technology. As defined for the project, a Research and Exploratory-Development Event (RXD) is a particular kind of increment. It consists of the occurrence of a novel idea and the subsequent scientific or engineering activity in which the idea is examined or tested.

The RXD Event was the basic quantifying unit in the HINDSIGHT study. It was the measure of new knowledge. In the evolution of a technology there might be one or two RXD events, or an extended chain of them. In any case, they would culminate in a new device or component found in a particular system or subsystem. RXD events were typically short in duration, lasting a few weeks to a few months; they might overlap, or they might occur sequentially. Summarizing the procedure of the study, the HINDSIGHT report notes the way in which each "differential amount of knowledge" was identified and then assessed in terms of its cost and its cost-saving.

> The teams of specialists identified 710 unique RXD Events, conducted the historical traces, and described and documented the related activities in terms of the differential amount of knowledge that accounts in part for the increased cost-effectiveness of the systems analyzed, compared with their predecessors (p. iii).

Both parts of each RXD Event, the novel idea and its validation, were precisely specified in the study. But the specifications boil down to two simple criteria. First, the criterion for affirming the novelty of an idea was simply the opinion of its immediate sponsors; it was novel if they thought it was. Second, the criterion for validation lay in some form of idea transfer, which the HINDSIGHT report refers to as "interface activity."

> By interface activity is meant simply the manner in which the knowledge generated by the first innovator is transferred to the second (p. iii).

Examples of interface activity include the preparation of a report, the presentation or publication of a technical paper, or a patent disclosure. Other interface activities include the demonstration of an idea's feasibility by mathematical analysis or by breadboard model, or attempts on the part of the innovator to sell the idea to a potential customer. They might even include an informal technical discussion among colleagues (p. 141f.). In sum, as a measure of knowledge production, each RXD Event is an instance of some conventional way of presenting an idea that is deemed to be novel, and of getting it validated by normal means in the R&D process.

Some of the findings of Project HINDSIGHT surprised the sponsors about the efficacy of scientific research in achieving technological outcomes. For example, the study revealed no support for the myth of the linear flow of knowledge from science to technology (p. xviii). But it did confirm the effectiveness of scientific research that focused on engineering problems (p. xvi). It also confirmed the efficacy of prior, specified requirements in advancing the development of a new technology (p. xviii). In any case, problem formulation was seen to be critically important (as we will see in a later chapter).

Although it is tangential to our present discussion about the measurement of knowledge, Project HINDSIGHT raised a historical controversy, worth noting, about the sources of such knowledge. At issue was the efficacy of independent basic research in technology development, and it provoked immediate contention from the science community. Project HINDSIGHT found *independent* basic research to be only a minor source of the knowledge increments that it described. Those findings prompted an NSF-sponsored project in 1968, called TRACES, which found to the contrary in the case of certain selected technological events. Then a follow-up study, "The Interactions of Science and Technology," sponsored by NSF five years later, confirmed the findings of Project HINDSIGHT. This second study found many more instances of major impetus to technological innovation from *mission-oriented* research than from independent basic research. (Zegfeld & Enzing, 1987, discuss these reactive studies.)

Overall, Project HINDSIGHT was an assessment of DoD's investment in technology development; it confirmed that costs could be tracked through the transfer of new technological information. These findings were immediately newsworthy. But the impact of HINDSIGHT lay also in the design of the project. Its pragmatic means of measuring knowledge enabled the use of such measurement to assess cost-effectiveness in a complex network of organizations.

The fact that knowledge could be measured meant that its value was quantifiable within a prescribed situation. That breakthrough also made it feasible to analyze the flows of knowledge from one point to another. The effect of such transference on the process of knowledge production is what we are about to explore.

4

The Dynamics of Knowledge Production

The Defense Science and Technology (S&T) Program centers on a canon of critical technologies with certain prominent characteristics. They are conspicuously oriented toward the processing of information, and they are largely interdependent. Like all technologies, they have knowledge content, and their development is so systematically managed that this knowledge is measurable by increments, as Project HINDSIGHT revealed.

How is such knowledge produced? The answer lies in some pattern common to the activities of R&D. Identifying such a pattern is the agenda of this chapter. The point of departure is fairly easy to locate in Project HINDSIGHT. Therein, the measurement of knowledge increments implies a sequence of some sort in the R&D process. But because this process is so open-ended, we need to explore it from several perspectives.

R&D is grounded in the idea of what might be called *directed discovery*. It is essentially an operational concept intended to expedite the management of many different discoveries that might possibly be related. But discoveries are continually made throughout the production and deployment of a technology, so R&D is really an extended set of undertakings, especially in the Defense S&T Program. This program is based on the premise that a technological device or system has a life cycle. The device or system evolves through numerous stages of research and development and also testing and evaluation; then it is produced and deployed and subsequently used. During its period of use it is maintained and upgraded until another device or system displaces it. Serving different local objectives along the way, the process of directed discovery occurs throughout the life cycle of a defense technology.

Moreover, this R&D endeavor is reflexive; it doubles back on itself. New discoveries in the evolution of a technological device or system often necessitate expensive revisions. So to minimize cost and maximize performance, DoD has progressively merged the development, production, and deployment of weapons. One effect of the merging has been to make R&D more open-ended by blurring the boundary between R&D and production. DoD has also blurred the boundary between production and deployment. To accelerate the transition of new ideas to their application in the field, the S&T Program calls for telescoping events that were formerly understood to be discrete in the R&D process. In Operation Desert Storm, the formal test-and-evaluation phase of certain new weapons was augmented by their actual testing in battle. And this operational emphasis has since been formalized. *Defense Science and Technology Strategy* (DDR&E, 1994) explains the current program for fielding new technologies through what are called Advanced Concept Technology Demonstrations (ACTDs). The program actively involves intended users in the operational testing of a prospective weapon while it is still in the stage of advanced development (p. 17f).

As evident from DoD's current management of technology development, the relationship of R&D to the life cycle of a technology is not at all precise. The process of managed discovery occurs continuously but in different ways at different stages in the evolution of a technology. For the purpose of analysis, subject specialists have isolated certain aspects of the whole process, including design, advanced development, and even innovation. To find a pattern of knowledge production in R&D, therefore, it is fruitful to look at each of these kinds of endeavor.

DESIGN

Engineering design is often associated with creativity. The American Society of Mechanical Engineers (1986) defines it as the creative process that proceeds from the identification of a need to a device or system which meets that need. Vincenti (1990) constructs a detailed, comprehensive explanation of this creativity. In *What Engineers Know and How They Know It*, he sets forth an explicit epistemology by representing the design process from three different perspectives. First, he describes levels of activities by which engineers typically formulate and solve problems. Then he relates types of knowledge to a set of knowledge-generating actions. And finally he offers a simple cognitive exercise by which engineers evolve new knowledge from previous experience. His three representations are all applicable to an open-ended notion of R&D.

Vincenti's first representation of engineering design is an algorithm, a set of five basic activity levels. These are stratified, each leading to the next, and the design process consists of working though them successively. This process model starts with *problem definition*, with the translating of some undefined or ill-defined requirement into a concrete technical problem. It proceeds to some representation of an *overall design*, a layout of arrangement and proportions to accommodate the project definition. It then proceeds to the *design of the major components* of the proposed structure. The last two stages entail *subdividing the areas of component design* and then *further factoring* those subdivisions.

For our analysis of knowledge production, the most important aspect of Vincenti's algorithm is what he does with it. Having set forth a neat hierarchy of successive events, he then muddies it up. "The complete design process then goes on iteratively," he explains, "up and down and horizontally, throughout the hierarchy" (p. 9).

Vincenti's second representation of engineering design is an anatomical model. He juxtaposes the kinds of knowledge that engineers use in design with the kinds of actions by which they generate them. He builds his model by degrees, first describing six kinds of knowledge (fundamental design concepts, criteria and specifications, theoretical tools, quantitative data, practical considerations, and design instrumentalities). Then he describes seven kinds of knowledge-generating actions (transfer from science, invention, mathematical engineering research, experimental engineering research, design practice, production, and direct trial). And he incorporates both kinds of values in a matrix. After suitable cautions about the inevitable overlaps among the values on both axes, he scores the matrix with check marks to show the kinds of cognitive interactions that occur. The matrix complements Vincenti's algorithm of problem formulation and solution. It explains how knowledge-generating actions dispose engineers to seek out different kinds of knowledge at different levels of the problem-solving hierarchy.

Vincenti's third explanation of engineering design is another process model. It is a personal, cognitive algorithm repeated over and over in the design process. Given any problem to be solved, the algorithm consists of three simple steps and as many iterations as needed:

1. Searching from one's past experience for novel variations that might serve the present need,
2. Compiling a list of those variations, and
3. Selecting the ones most likely to work.

In our search for patterns, several inferences from these models are worth noting. One is struck, for example, by how generalizable they seem to be. The hierarchy and the random iterations that characterize Vincenti's problem-formulation model extend well beyond engineering to broader use in formulating problems. And as for his cognitive model about searching one's past experience for variations that might work, who has not prayed at this altar?

A different kind of comfort can be had from Vincenti's anatomy of engineering design; this neat, static model cites seven different kinds of knowledge-generating actions that characterize design. Looking back at those actions, however, we see that two of them—production and direct trial—go far beyond the conventional boundaries of R&D; they stretch well into innovation. This extended reach suggests that the design process is endemic to all of technology development; it suggests that knowledge production is holistic.

One other notable characteristic of the problem-formulation process is that it occurs primarily in a social context. Vincenti describes an astonishing number of technical specialists who are affiliated with one another in formulating and solving problems; by turns, they refine the knowledge that they pass from one to another. Through the nature of their work they are community-oriented. Their communities are formal research institutions—commercial, educational, or governmental—as well as informal groups of specialists assembled for particular tasks in the course of design and production. Both competition and cooperation in these communities reinforce individual learning. Group activities also provide the occasions and the means of transferring what is learned from one organization to another. Notwithstanding the cherished myth of the solitary inventor, engineering design is typically a collective endeavor.

ADVANCED DEVELOPMENT

The Defense S&T Program partitions development into two successive stages, exploratory and advanced. Exploratory development includes much of the design process that Vincenti describes; it is expansive. Advanced development, on the other hand, focuses more narrowly on the manufacture of a device or a system that has already been designed; it has been characterized as that span of activities between the developer's decision to commit resources to manufacturing a product and the actual manufacture of that product.

This characterization appears in a series of papers (Thomas Marschak, Thomas Glennan, and Robert Summers, 1967), entitled *Strategy for R&D: Studies in the Microeconomics of Development.* The papers were prepared at the Rand Corporation and funded by the Air Force at about the time DoD undertook Project HINDSIGHT. From their different perspectives, both efforts address cost-benefit analysis as measured by knowledge increments.

The Rand papers begin by declaring that development is a process of combining new knowledge with old knowledge to achieve new and useful products. This statement is then narrowly interpreted to mean that development is a process of reducing uncertainty. As described in the first Rand paper, the process is reactive; it consists of responses to knowledge already acquired about the device that is to be manufactured. Successive increments of knowledge about the device entail a sequence of responses that reduces uncertainty.

So far, this sounds like a fair representation of what happens in basic research. But the uncertainty that marks the advanced development of a technology is a legacy of earlier, exploratory research about the device; it represents risk, which must now be either eradicated or minimized. Risk typically becomes evident when a laboratory prototype is moved to a more stringent set of operating conditions prior to production. Such is the risk that Gansler (1989) speaks of in explaining the critical leverage of research over the later stages of developing a technology.

The second Rand paper describes the importance of economic modeling in the reduction of uncertainty. The purpose of modeling is to identify the important determinants of cost, time allotment, and other issues that bear on the later production of a device. Modeling quantifies the uncertainties inherited from earlier stages of the R&D process. Because they cannot all be resolved before production is begun, they pose a risk which needs to be resolved by making the best trade-off between maximum performance quality and acceptable cost. In any case, the magnitude of uncertainty depends partly on the constraints of time and money allowed for development and partly on the extent to which a new technology is intended to advance beyond the current state of the art. The greater the advance, the greater the uncertainty. And the amount of risk is always a function of how far the developers are venturing beyond what they know.

The magnitude of the risk determines the magnitude of uncertainty. But uncertainty is not ignorance. On the contrary, it is one kind of knowledge; it is the knowledge of what yet needs to be learned. Here is a tight, focused view of knowledge production, more confined than what emerges from Vincenti's perception of design. This view of advanced development presumes that knowledge production consists of making controlled responses to changes in existing knowledge.

INNOVATION

Yet another perception of knowledge production emerges from the perspective of innovation. A comprehensive concept, innovation is by far the broadest way of designating how a technology evolves and becomes operational. Tornatzky and Fleischer (1990) describe it simply as "the process by which new tools and devices are developed and introduced" (p. 24). This loose description covers most uses of the term that are embedded in a large literature on the subject of technological innovation. Commentators in different disciplines tend to appropriate the term in its entirety, even when they are examining some specific aspect of it. (Some examples of this usage are in Arthur Gerstenfeld, 1977, in Wesley Shrum, 1885, and in Michael Tushman and William Moore, 1988.) In its various usages, the term reaches backward to exploratory development and forward to deployment and marketing. But whatever its reach, the term *innovation* is typically grounded in the marketplace. One durable definition, in *Government Involvement in the Innovation Process* (MIT, 1978), specifies that "Innovation is the *commercial* introduction of a new technology." As might be expected, it is not a linear process.

> The innovation process includes a complex and interconnected set of activities leading from invention to commercial introduction, but not necessarily in any prescribed sequence (p. 11).

Innovation is a slippery concept, given its long reach from early R&D activities to the ultimate application of a technology. Yet its comprehensiveness accurately conveys the scope of DoD's engagement with technology. The key to managing innovation, according to Tornatzky and Fleischer, lies in recognizing that "developer and user are not two isolated phenomena, each with its own language and issues, but rather coparticipants in an intricate dance, the object of which is the enhancement of both parties" (146). To extend the metaphor, I might add, DoD represents both partners in the dance; it both leads and follows, first sponsoring the development of the technology that it then implements.

DoD's style of innovation, as I mentioned, is based on the notion that a technology has a life cycle. This notion influences the way in which knowledge moves from one group to another. In reviewing the literature of innovation, Tornatzky and Fleischer point out the main features of the system life cycle (SLC) concept, which originated with software developers. Not surprisingly, these features all match those of the defense critical technologies. SLC reflects the heavy emphasis in software design on meeting some particular set of prior requirements; from the outset, users' needs are built into the system. SLC's other features are derived from this premise: the importance of connecting each system to other existing systems, of early planning for inserting the system into an ongoing operational environment, and of maintaining the system after it has been put into use. These features are prominent in the defense critical technologies.

Wendy Faulkner (1994) has tracked the different uses of knowledge that innovation entails. Her purpose is to learn more of "what happens to knowledge as it moves between, and is developed by, different groups in the course of innovation" (p. 426). She finds that the cumulative nature of technological development is a recurring theme in the economic histories of technology (p. 440). Concentrating on private sector firms, she remarks on the importance of innovation to knowledge flow among companies in the supply chain: from subcontractors to contractors, from contractors to customers, and also between competitors. Informal interaction is typically the means of maintaining this knowledge flow. Faulkner's findings about the commercial marketplace are also true of the defense establishment. As we are about to see, the Pentagon presides over this same kind of interaction in the procurement of its technologies.

To schematize the kinds of knowledge that move from one place to another, Faulkner incorporates the models of Vincenti and others into a larger construct. She builds a composite typology of 15 kinds of knowledge in five categories, complete with refinements and qualifications. Then she explains the reason for all the overlapping categories, the blurring of discrete types, and the fluidity of it all:

> This is the nature of the beast: knowledge used in innovation does not come in watertight boxes but is mutable and multidimensional, precisely because of the complex social processes by which it is generated and utilized (p. 449).

PATTERNS OF KNOWLEDGE FLOW

Although they are described from different perspectives, these several phases in the life cycle of a technology have something obvious in common. Diverse though they are, design, development, and innovation are all community endeavors. As regards knowledge production, they reveal an interplay of initiative and response. In all three aspects of R&D, knowledge moves from one locus to another, within groups and between groups, and responses to knowledge produce new knowledge. These evident factors in a pattern of knowledge production were anticipated in two seminal studies reported by Thomas Allen (1977/1986) in *Managing the Flow of Technology*. Allen's studies reveal the patterns of communication among scientists and engineers in the aerospace and electronics industries. Conducted in 33 laboratories engaged in defense R&D, the studies describe the problem-solving behavior of scientists and engineers during their exploratory research. Their activities consisted mainly of preparing feasibility studies and design studies, and occasionally of building prototype devices. The work occurred in essentially an engineering community, and it reflected community interaction. As Allen observes, engineers typically work in groups; organizational teaming predominates, and the teams are mission oriented. They work toward the achievement of assigned objectives, and they are obligated to produce deliverable knowledge products. And, as Vincenti has explained in more detail, their work is problem oriented; it is iterative: repetitious and incremental.

The acquisition of information in this community occurs largely through personal contacts. In addition to exhaustive data tables, the literature that is sought comes mostly from textbooks, proprietary reports, trade publications, and less frequently from research articles in learned journals. Within the laboratories, the exchange of information is largely informal. Members of task groups consult their colleagues in other disciplines as needed. Such activity typically concerns the generating of criteria for problem formulation as well as for defining acceptable solutions. A significant amount of communication occurs between specialists in different laboratories. Although the laboratories typically compete with each other, they all have a stake in maintaining a common knowledge base in the fields that engage them.

No R&D laboratory is entirely self-sufficient. As Faulkner (1994) has also observed about knowledge flow among commercial firms, the companies in the supply chain typically communicate within fixed channels: subcontractors inform contractors, contractors inform customers, and competitors share what is mutually beneficial. Other agents of sharing are independent consultants, who typically serve more than one client. Vendors are important communicators. Allen (1977) observes that they are important potential suppliers of components and subsystems, and that they provide information that they hope will stimulate future business. He then generalizes about the importance of customers and vendors in channeling the flows of information within the R&D community.

> Involvement in the marketplace, either through the customer or potential vendors, exerts a significant influence upon the communication system, providing channels of information in two directions and connecting buyers and sellers through both the procurement and marketing functions of the organization (p. 45).

The key words in this portentous statement are *marketplace*, *customers and vendors*, *procurement*, and *marketing*. They appear to be the factors that decisively affect knowledge production in defense science and technology.

Although Allen's studies focus on the early, exploratory stage of R&D, these economic factors are germane to the entire range of knowledge-producing events in the evolution of a technology. They are elements of a pattern that Fritz Machlup (1980) has explicated in *Knowledge and Knowledge Production*.

Machlup's premise is that knowledge is a marketable commodity, and that it is exchanged, like other goods, between customers and vendors. But there is an extraordinary quality about this commodity. Unlike other goods that come to market, knowledge is actually produced by the exchange process. Many times repeated, the buying and selling of knowledge is the basic pattern by which knowledge is produced in the enterprise of defense science and technology.

Buying and selling knowledge produces knowledge. This startling proposition appears to be self-contradictory, but like all paradoxes it can be explained by valid deduction from acceptable premises, which Machlup does. In the course of his explanation, moreover, he demonstrates that exchanging knowledge not only produces knowledge, but that it is also a means of measuring it.

A MARKET PATTERN OF KNOWLEDGE PRODUCTION

These various perceptions of knowledge and its dynamics lead us to an implicit pattern of knowledge production in the defense R&D community, and Machlup (1980) provides language to describe it. Knowledge—that is, what is known—has economic value. Its value lies in its being unique or innovative, or in Machlup's words, "socially new." He uses the phrase to mean original knowledge, that which is known in its earliest instance, either universally or personally. What one knows for the first time is socially new, whether or not it is already known by others. With this inclusive view of knowledge, he defines its production as "any activity by which someone learns of something he or she has not known before, even if others have" (p. 8).

Socially new knowledge has the potential of being bought and sold. R&D is particularly marketable, since it can be regarded as an investment that will pay off in future productivity. "In the case of consulting firms, independent research-and-development organizations, and other information services, descriptions in terms of supply and demand may be very much to the point" (p. 184). This is the frame of reference that we will chiefly use from here on to explore the dynamics of knowledge production, so it merits more explanation.

The commodity value of knowledge is the premise of Machlup's theory that knowledge production can be understood in the context of an economic market; it can be described, measured, and assessed in market terms. "Reference to a marketplace implies the presence of two groups," he observes, "one offering knowledge for sale, the other desiring to acquire it and willing to pay for it" (p. 183f). The vendor and the customer and the saleable goods are the components that make a market, and the stocks and flows of those goods are its dynamics. In these basic respects the market for knowledge is like other markets.

> In economics a fundamental distinction is commonly made between stocks and flows, usually with reference to goods, to capital funds, to money. The distinction also applies to knowledge. At any moment of time, there is a stock of knowledge; during any period of time, there is a flow of knowledge (Machlup, 1980, p. 161).

Machlup builds on this proposition. As with the passage of any other goods through the marketplace, the passage of knowledge from one party to another is characterized as a stock-flow-stock relationship. The stock of goods flows from one party to another.

But Machlup makes an important distinction. Unlike the flow of other goods, the flow of knowledge from one party to another increases the combined amount of both stocks. The explanation of this paradox is that knowledge exchanged from one party to another now belongs to both parties. The transmitter retains whatever knowledge the receiver gains, so the combined stocks of knowledge are increased.

To assess the magnitude of the knowledge increment, an economist needs a way to measure it. Machlup addresses this subject as well. Studying the stock-flow-stock relationship, he observes from the literature of various information services that knowledge is generated, disseminated, and used. He also observes that the generation of knowledge cannot be readily quantified, except perhaps in a researcher's exhaustive laboratory notebook, which is not easily accessible. Neither can the use of knowledge be quantified. He cites, for example, the difficulty of trying to measure the adaptation of some new technological idea by tracking the countless litigations over patent infringement.

Both the generation of knowledge and the use of knowledge, in Machlup's words, are "nonoperational concepts"; that is, they are not statistically accessible for measuring the social value of knowledge. Only *dissemination* is useful for this purpose (p. 175). Flows of knowledge can be quantified and assessed "by the measuring rod of money," as he puts it. To summarize Machlup's several sets of assertions, knowledge flow augments the stocks of knowledge, and it is also the means of measuring how much knowledge has been produced.

KNOWLEDGE FLOW AND KNOWLEDGE PRODUCTION

Earlier, when we were coming to terms with information and knowledge, I proposed that knowledge is a species of information or, more accurately, a unique rendering of it. Knowledge is derived by a cognitive process of interpreting information according to some need to know, some schema. The schema might be *a priori*, or it might evolve as one interacts with accessible information. Either way, the interpreting of information is the essential knowledge-producing process.

The relevance of this model to Machlup's sense of knowledge acquisition in the marketplace becomes apparent. The flows of socially new knowledge, he says, are governed partly by the intentions of knowledge buyers, of whom there are three sorts. One is the diligent knowledge seeker who is prepared to incur costs to find new production techniques, product innovation, marketing, or some other response to the market. The diligent seeker looks for answers to questions that may be vaguely circumscribed or highly specific. A second kind of potential customer is not deliberately seeking knowledge but is alert by nature and open to acquiring useful knowledge at little or no cost. A third potential buyer is the accidental finder who stumbles across some knowledge as yet undiscovered but potentially useful. To these three instances of knowledge acquisition Machlup adds another, the fortuitous event; this is the unintended discovery of one kind of knowledge by the diligent seeker who is searching for another. For this instance, Machlup invokes the concept of serendipity.

Each of these seekers, we might add, already has some intention, some need to know, which makes possible the recognition of relevant knowledge when one sees it. But intentions account only partly for the flows of knowledge through the marketplace. According to Machlup, they blend with the initiatives which potential buyers undertake, "especially institutional devices or governmental measures designed to increase the inducements to produce knowledge" (p. 183). In such cases, knowledge stocks are disseminated in response to demand. But initiatives can also come from the supply side of the market; knowledge providers may increase stocks of knowledge in the expectation that there will be a demand for it.

Given the continuous reciprocity between latent supplies and demands, it frequently occurs that one creates the other. Potential buyers induce suppliers to increase their stocks of knowledge, and suppliers create demands for those stocks, as, for example, in the business of grants and contracts. With this interactive situation, Machlup has effectively explained a marketing environment, one that will become increasingly significant in our exploration of knowledge production.

RHETORIC AND ECONOMICS

The exchange of knowledge as a market transaction is an elegant idea; it conveys so much so precisely. It also clarifies the complexities of defense science and technology by grounding them to an accessible pattern. Fundamental to economics, the idea of a market driving many disparate energies provides a durable model for exploring knowledge production.

The dissemination of knowledge produces knowledge. That is the essential proposition. Its continuous repetition in the enterprise of defense science and technology is the pattern of knowledge production. In terms of economics, the exchange of knowledge between customer and vendor increases their combined stocks of knowledge because knowledge survives the transaction; it remains in the knowledge stock of the person who transmits it as well as augmenting the knowledge stock of the receiver.

Yet there is more to the proposition than economics by itself can explain. The exchange of knowledge is by nature generative. And we are back to cognitive process again. Knowledge exchange is a communicative act wherein the sender's own stock of knowledge actually *increases* through the act of communicating it. This phenomenon is not merely the arithmetic increase of duplicating one stock of knowledge by transferring it to another. It is of a wholly different order, more like what occurs in the Biblical episode of the loaves and the fishes. The exchange of knowledge actually produces knowledge that did not exist until the communication occurred.

Here is a startling marriage of rhetoric and economics, yet the generative quality of communication is commonplace. It is as common as the composing process, which occurs everywhere all the time. Composing is a discovery process so familiar that we take it for granted. The discovery is as commonplace as writing a letter and then deciding not to send it or to use the telephone instead. The act of composing that letter generates some new perception that resolves whatever was unwieldy. Composing happens everywhere around us. The person in the audience who rises to ask the speaker a challenging question and makes a rambling speech instead is engaged in composing. The composing process entails continuous revision, such as explaining the directions for how to get from here to there and then amending them along the way as new realizations occur. Composing accounts for that realization, "I open my mouth to hear what I think."

Explaining is one way of persuading, which is the essential characteristic of rhetoric. Persuasion in any verbal mode entails composing, which in turn entails discovering new relationships en route. Because composing is generative, so is persuasion. This proposition adds rich dimensions to Machlup's steadying structure of an economic market wherein customers and vendors exchange knowledge for money. Machlup's structure underlies the rest of this study. But the theme that it supports pertains to rhetoric; it is about persuasion, about defining and also creating the kinds of knowledge that are marketed for sale or purchase.

Economic structure and rhetorical theme work together here. The structure reveals the underlying pattern of knowledge production in the R&D community, wherein the dissemination of knowledge produces knowledge. The resonance of the theme lies in the myriad repetitions of the underlying pattern and in the disparate energies that the market coordinates. Let us explore them together: the economic structure of the market and the pervasive rhetorical theme of persuasion.

Coming to Terms
II

The next four chapters deal with the defense R&D community. Converting information to knowledge in this community is a function of people interacting; an adage one hears is that "R&D is a people business." But those personal interactions are channeled by organizations that transact with one another in complex, and sometimes bewildering, ways.

Patterns of knowledge production in the defense R&D community get blurred by the immensity and the complexity of its organizational context. What comes to mind is a gestalt, a visual figure whose properties cannot be derived from its parts. If you stare at it long enough it takes another shape; foreground and background change places, and perspective disappears. Tracing the flows of knowledge in this community provokes a similar sensation. They remind one of M. C. Escher's drawings, in which skewed stairways at different elevations merge with one another, and water sometimes flows uphill.

Coming to terms with an Escher drawing sooner or later entails studying the context and trying to see how the contrary details are lodged within it. So it is with the defense R&D community. It functions in the context of an economic market, but it is unique within that context. Chapter 4 addresses this peculiar relationship, noting how the federal government regulates both customer and vendor in the defense technology market. The major customer is the DoD, and its role in the market is described in chapter 5. Actually, this customer is an infrastructure of discrete organizations defined by their respective missions. Those intramural organizations sell their

R&D services to one another, as well as buying them from extramural vendors. Chapter 6 describes the extramural vendors. It points out the surprising array of relationships that they formalize with their customers and with one another. Both within and beyond DoD, customers and vendors interact in a marketing environment. The ways in which marketing functions in those interactions is the subject of chapter 7.

The sequence of these four chapters might be called cinematic. It begins with a wide-angle shot of the whole scene, the whole defense market in its historical context. Then the ensuing chapters focus on successively narrower scenes: on the customers and the vendors, and then on their ways of positioning themselves. The sequence accumulates more and more interactions along the way, anticipating a tight shot of how the participants do business with one another.

THE MARKET

Since "market" is central to this set of chapters, we must come to terms with it. But the market for defense science and technology is peculiar. The forces that drive it are political as well as economic, and they are sometimes at odds. Moreover, they are exerted in the form of paralyzing regulations. Understanding how this uncommon market evolved depends on recognizing its origin. The idea of a free economic market, conceptually and historically, is our point of departure. So we first need to come to terms with that, and then with the way in which the defense technology market differs from it.

Although we usually refer to a market as a single entity, it is more accurately a collective term, one that represents many kinds of interaction. To define it in a single predication is a *tour de force*. But since we need a place to start, here is a beginning:

> A market is an aggregation of parties engaged in exchanging goods under certain limiting conditions.

The point of the definition is not to distill the term so much as to open it up, since nearly every element in it needs to be qualified or expanded. The goods are properties or services of one sort or another; the parties engaged in the exchange are customers and vendors. The syntax of this definition features the involved parties. They may be individuals or groups or corporate entities or governments, and they are of two kinds. They either own or control those goods that they offer for exchange, or they want to purchase them with money that they own or control. The offerors are vendors; the purchasers are customers. In marketspeak they are suppliers and demanders.

The conditions that limit the exchanges may include regulations imposed by government. The principal constraint on the exchange of goods in a free market, however, is not external. It is internal; it inheres in the exchange process, which is what makes the market work. To emphasize this process and its inherent constraint, it is convenient to revise the syntax in the definition of a market that I offered a moment ago. That definition centers on the interacting parties, the suppliers and demanders. But featuring the exchange process instead helps to clarify where this crucial constraint is lodged. So here is the first definition, turned around:

> A market is the aggregate exchanging of goods among parties under certain limiting conditions.

The principal constraint on a free market is inherent in the exchange process. The constraint is a relationship among three different values: the supply of the goods offered for exchange, the demand for them, and their consequent price. As the relationship changes between supply and demand, so does the price. I mention this obvious relationship because it is the first respect in which the market for defense science and technology departs from normal market characteristics. In the defense S&T market, price is not necessarily fixed by supply and demand.

The distinction between the two kinds of market can be graphically displayed. The dynamics of a free market, such as Adam Smith described in the 18th century, is usually represented by graphs in microeconomics textbooks. Therein, three variables operate in a closed system. Supply and demand are independent variables, and price is dependent on their relationship. If the demand for goods remains constant, their price moves up or down with the available supply. Similarly, if the supply of goods is fixed, their price moves up or down with the demand for them.

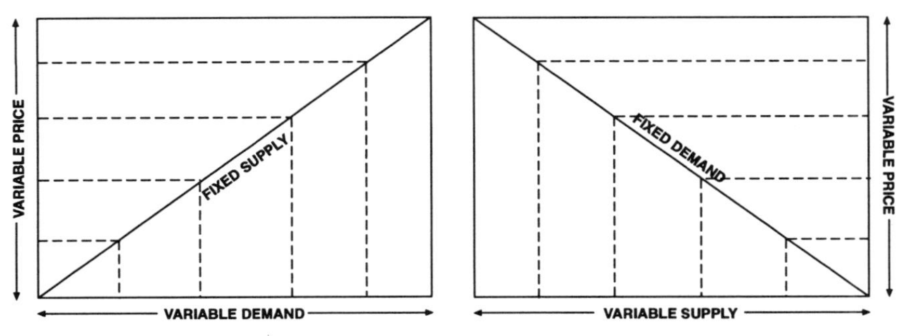

FIG. II.1. The dependency of price on supply and demand.

The logic lies in the mathematics. Any intersection of supply and demand yields a corresponding price. But in the defense S&T market, supply and demand do not particularly matter in the price of the goods except in random instances; price is largely a function of other factors. On either of the above graphs, for example, the price of goods in this market would hover somewhere in space, without reference to any point where supply and demand intersect.

In the defense S&T market, the price of knowledge-based products is largely a function of the costs of producing them. Such costs result from a variety of political causes, some of them laudable. They result, for example, from expensive regulations about protecting the environment or the interests of small business. Or they can result from "pork", from stipulations that R&D services be performed at designated facilities within some particular congressional district without benefit of competition. The price of goods is also a consequence of a peculiar culture within the defense acquisition system. This culture tends to favor initially low cost estimates followed by cost overruns after production is underway. Moreover, price is always subject to a federal cost-accounting procedure that distributes fixed costs evenly over short production runs, which accounts for those infamous $200 screwdrivers and $800 coffee pots.

These kinds of costs are embedded in contracts for R&D. Stories about them abound in the R&D community, all of them told by people who do not want to be quoted; they document the illogic of defense acquisition policies and procedures. The stories tend to be local, but you can get the general idea of them from the overstatements of *Augustine's Laws* (Augustine, 1983), which has Biblical authority in the R&D community. Formerly a high level manager of R&D acquisition in the Pentagon, Norman Augustine is now CEO of Lockheed Martin, probably the nation's largest defense contractor. He explains that the process of generating requirements for future U.S. weapon systems "basically consists of enthusiastic engineers in industry promising to deliver 'all the capability you want for a dollar' and a customer which responds 'I'll take two dollars worth.' It seems that some people just won't take 'yes' for an answer" (p. 44).

Jacques Gansler (1989) comments more soberly on the same phenomenon. In *Affording Defense*, he points out that the acquisition of weapons systems actually has an exceptionally low incidence of errors or abuses; such excesses are remarkably infrequent compared to those in any other private or public R&D operations in the United States. What he argues, rather, is that the high cost of production is already designed into weapon systems by prior military specifications for high performance; these MILSPECS are built into the R&D process. They cause over-complicated designs that make manufacturing difficult; assembly time is lengthened, and rejection rates are high, resulting in fewer weapons at higher unit cost. MILSPECS not only drive up production costs, they also necessitate unanticipated maintenance costs after a weapon is deployed.

THE CUSTOMER

In addition to indulging, even encouraging, high costs, the market for defense science and technology has other features that help make it unique among markets. One of them is the unusual nature of the customer; we must come to terms with it. The constituent organizations of DoD make their purchases through a comprehensive structure of planned events known as the defense acquisition system. The events begin with a needs assessment, and they include the procurement and then the disposal of whatever remains of the purchase. The system behaves like any individual consumer, so much so that we are likely to overlook the obvious. DoD *is* a consumer; it is the end user of virtually everything it buys. This fact distinguishes DoD from other federal agencies that sponsor R&D for the benefit of their various constituencies in both the public and private sectors.

DoD is the largest consumer organization in the world. Its annual acquisition budget exceeds the total budgets of many foreign governments, and its organization is vast.

Reporting for the OTA, Jack Nunn (1994) states that DoD's acquisition workforce exceeds 178,000 persons, at an annual cost of $7 billion in salaries. (To communicate formally with one of them, and to expedite an answer, you address your correspondence not to the individual but to an office number which designates the function of that office in the larger organization.) This labor force represents a comprehensive knowledge base so specialized in acquisition that other federal agencies commonly ask DoD to make purchases for them: a helicopter for the Coast Guard, perhaps, a computer network for the Merchant Marine or the Federal Aviation Authority, or some laboratory equipment for the U.S. Coast and Geodedic Survey. DoD can readily make such designated purchases because it has immense purchasing power.

In most of the remarks that follow, it will be convenient to use the term "customer" in a generic sense. I will use the term to designate any DoD organization that buys R&D; to be consistent I will do the same with the term "vendor," to designate any supplier of R&D in this market. Although convenient, this usage is not always precise, since each of the discrete R&D customers is shaped by its own mission and its own organizational culture. On the whole, however, the purchasing procedures of all these R&D customers are centralized by the defense acquisition system and constrained by federal regulations that govern the procurement of all goods, including R&D. From a vendor's perspective, these glacial forces tend to make the R&D customer generic. They homogenize purchasing procedures, and they result in aggressive auditing. In the case of R&D purchases, they also entail on-site monitoring to control the way vendors do business. Moreover, the government owns some of the facilities, notably laboratories, from which it buys its knowledge products. These market features that arise from the unusual predominance of the customer become evident in the ensuing chapters.

SCIENCE AND TECHNOLOGY

So much for "market" and "customer." The usage of one other significant term, "Science and Technology," represents some unfinished business and needs also to be addressed. As you recall from chapter 2, the term is used in parallel ways at the Pentagon. Conceptually, it designates a knowledge base; programmatically, it denotes the organization of resources assigned to develop that knowledge base. The usage is slippery, since the official language does not always clarify which referent is intended. Even more, the phrase itself is ambiguous; it is accurate as far as it goes, but it is incomplete. Science, meaning fundamental science, and technology are joined in the same knowledge base, whether we think of them as process or as output. But the inclusive term does not indicate any relationship between its two parts. Official usage observes the linkage of science and technology only in the programmatic sense, utilizing budget categories to imply what part of the knowledge base is science and what part is technology. The ambiguity is probably desirable, but the schema tends to encourage a more casual usage that randomly conveys either of two ideas: that science and technology are one and the same, or that they are linear. In fact, they are neither.

Had I been writing this 30 years ago, I would have been trying to delineate the two domains in their presumed sequence. But recent scholarship has fairly settled that matter. The linear model of progression from science to technology is an idea whose time has gone; it was seen to be "fundamentally flawed," as Faulkner (1994) observes in her review of the typologies of technical knowledge. She cites D. J. de Solla Price (1965), who had earlier remarked on the discrete nature of the two entities, with science tending to build on old science, and technology on old technology. What has replaced it, Faulkner (1994) finds, is the concept of "a strong interactive relationship between two semiautonomous activities" and a blurring of the boundaries between them (p. 427ff). This is surely the case in the development of defense science and technology, as the next few chapters show. But the myth of linearity is durable. As it pertains to science policy, Richard Barke (1991) discusses the curious persistence of the linear model that lingers in congressional circles. The promise of science leading to technology, he notes, has an appeal that helps to justify the funding of science programs.

The relationship between science and technology is not simply ideological. In the United States, at least, federal funding is the key to it. In *Lost at the Frontier*, Deborah Shapley and Rustom Roy (1985) vigorously attack the habit of polarizing basic science and technology, especially in matters of funding. They would restructure basic science as it is now institutionally conceived and replace it with a more coherent scheme for sponsored research. The scheme they project resembles what Vannevar Bush (1945) in *The Endless Frontier,* originally conceived as a national research policy. Specifically, they propose that sponsored research be restructured so that it can be funded in five different categories. The proposed categories are (a) undirected basic research (which they see as now dominating all scientific research), (b) purposive basic research, (c) applied science, (d) engineering, and (e) technology.

There is a point to be made here about the nature of defense science and technology. Shapley and Roy have written a reasonable, deliberate polemic, and they freely acknowledge the immensity of the social task that such a change would entail. Herein lies the point. Something very close to the radical realignment of science funding that they propose has been operating within DoD since the 1950s. The development and maintenance of DoD's technology base incorporates all five categories of science and technology that Shapley and Roy propose. The Pentagon is a house of many mansions, and the Science and Technology Program is one of them.

ACTIONS, ACTIVITIES, AND OTHER MATTERS

In following the flows of money, there are a few other technical terms to be noted. They present no conceptual challenges, but they do depart from common usage, deliberately and sometimes arbitrarily. Such is the case with the terms *actions* and *activities*.

In the defense S&T Program, contrary to one's expectations, these two words are not at all similar. An "action" is a certain kind of transaction, one that results in a contract or a grant, or in the modification of either instrument; the term is generally used for statistical purposes. In a given R&D program, for example, we might see that there were 1,134 actions last year. An "activity," on the other hand, is an organization. Money appropriated by Congress, for example, is distributed to the activities in the R&D infrastructure. Activities have missions, in support of which they both purchase and perform R&D. And their R&D purchases are tabulated as actions.

Accommodating this abrupt shift in meaning does not come naturally; it takes some practice. Yet the usage is stable, even though the number of activities might change from year to year. *The Department of Defense In-House RDT&E Activities* is an annual roster issued retroactively by the Office of the Secretary of Defense (OSD). For FY 1990 (OSD, n.d.), for example, it listed 95 in-house activities. About half of them were Army activities; the other half were about equally divided between the Navy and the Air Force. The total also included a few activities reporting to OSD. As a result of the military drawdown, however, the roster for 1992 (OSD, 1994) lists only about two-thirds as many activities. Some had been eliminated or combined with others.

A postscript may be in order here. Two other terms, in passing, are idiosyncratic and interesting. Both terms convert nouns into verbs. One is the word "transition." For example, "The objective of this project is to transition the prototype to the field for further testing." The other converted noun is "task," which is shorthand for assigning a task or a responsibility. For example, "In this project we will task the Cold Regions Test Center to transition the prototype to the field within 60 days." Both of these nouns become efficient verbs. If you are an insider in the R&D community, you probably will not think them unusual; if you are not, then their usage might seem ungrammatical. But either way, language belongs to the people who use it.

5

Heavy Hands on the Market

Herman Melville has dramatized the awesomeness of high stakes and high risk, in a tiny, symbolic chapter of *Moby Dick*, called the "Lee Shore." Therein, Ahab's ship, the Pequod, sets out to sea in a storm. Facing fierce headwinds that threatened to dash it back onto a rocky shore, it can only move seaward; Confirming Ahab's vision, it must seek whatever safety there may be in the howling, infinite landlessness. From the deck where he crouches, braced against the wind, Ishmael realizes that heading out to the infinite unknown is the only alternative. And he perceives this fearsome choice to be a metaphor of the human predicament.

The plight of that symbolic ship suggests the situation of the defense R&D enterprise, hovering at sea between two powerful forces, the market and the government. Each one threatens potential dangers. Aboard the Pequod, where one vision prevails, the dangers are more clearly defined. But they are much less so for the R&D enterprise, in which critical decisions are typically subject to disparate points of view and rancorous disagreements. Distributed among several sources of political power, the vision of defense technology is often diffracted. The ways in which those political forces affect the defense R&D enterprise is our subject.

The process of knowledge production in the Defense S&T Program begins with congressional funding. Article 1, Section 8 of the U.S. Constitution, which empowers Congress to call forth the militia to suppress insurrections and repel invasions, implies the funding authority. And the appropriation of money is inevitably subject to congressional strictures on how the money will be spent. Such strictures, in many forms, establish the basic conditions of the defense S&T market. They are embedded in the definition that I offered when we were coming to terms:

> A market is the aggregate exchanging of goods under certain limiting conditions.

Passed into law and interpreted through many different visions, these limiting conditions have grown enormously during the past 50 years. They are reflections of how Congress chooses to use the need for national defense to serve other purposes as well.

Governmental intervention in the market has a history that helps to explain the defense S&T market. This chapter provides a bifocal view of how this market works, and how it came to be what it is. Like virtually all modern markets, this one has evolved from the classical model, *An Inquiry Into the Nature and Causes of the Wealth of Nations* (Adam Smith, 1776/1976). We turn first to Smith's theory of the market, with particular reference to the issue of free exchange versus state regulation. Some interpretations of this issue then lead us to the formidable system of authorizations and regulations that govern the acquisition of United States defense technologies.

ADAM SMITH'S MARKET ECONOMY

The Wealth of Nations has the authority of a secular bible. More than 1,000 pages long, modified through five editions in Smith's lifetime, his treatise was a work in progress for some 25 years. In the economic analysis of markets, it remains a major point of reference. For more than 200 years, in the light of new political events, economists and economic historians have reassessed elements of Smith's plan for a market-based economy. They have kept asking, "How well did he explain what is happening today?" And their interpretations of his plan amplify its authority. As Jacob Viner (1928) has pointed out, "traces of every conceivable sort of doctrine can be found in that most catholic book, and an economist must have peculiar theories indeed who cannot quote from *The Wealth of Nations* to support his special purposes" (p. 126). The Smith industry, like the Darwin industry, has kept the original work current by focusing on what is problematic about it.

Wealth describes how human beings behave when they engage one another in trade. It anticipates inequities, raises cautions, and proposes guidelines for dealing with the excesses that trading naturally engenders. Part of its legendary quality lies in Smith's authoritative voice. He spikes his discussions with axiomatic opinions in a manner typical of its times. Its style recalls the trenchant judgments in Samuel Johnson's dictionary and Benjamin Franklin's aphorisms in *Poor Richard's Almanac* Like those contemporary writings, *Wealth* represents the 18th-century confidence in sweeping pronouncements about the moral nature of humankind. During the Enlightenment it was politically correct to generalize the motives and actions of all people from a few conspicuous instances. Even the making of such public pronouncements confirmed their authority; because they were uttered, they seemed true. So there is a lingering habit among noneconomists to assume that Smith's treatise explains the market as if it were a fixed entity. One of its legacies, therefore, is the mistaken notion that all markets are alike.

Smith graduated the human habit of exchanging goods into an entire economic system. By symbolizing self-interest in terms of money, he invoked a common denominator that virtually every person could comprehend. Charles Bazerman (1993) offers a rhetorical explanation of this linkage, pointing out that Smith's stunning invention was to correlate money with human desire. Smith made this powerful symbolic relationship the basis of his economic system, thereby establishing broad acceptance of the system itself. Its dynamics lay in the free exchange of goods for money.

Smith's model of a money-based economy assumed a natural price for goods based on several variable elements: cost, supply, demand, and competition. Depending on these variables, market prices would change, thereby accommodating any surplus of goods. An equilibrium among them would enable the economy to grow and develop. Smith also acknowledged one other market element, one that pertains particularly to the U.S. defense R&D market. That element is the regulation of the exchange process by the state, to counter monopolies or other arbitrary barriers to fair trade. Closely related to its regulatory role in Smith's model, the state has three duties to perform. In different ways, they all pertain to the subject of knowledge production for the public welfare. Book IV of *The Wealth of Nations* concludes with a summary of these duties. Here is that passage, edited to its essentials:

> According to the system of natural liberty, the sovereign has only three duties . . . of great importance . . . plain and intelligible . . . to common understandings:
>
> first, the duty of protecting the society from the violence and invasion of other independent societies;
>
> secondly, the duty of . . . establishing an exact administration of justice; and,
>
> thirdly, the duty of erecting and maintaining certain public works and certain public institutions, which . . . can never repay the expence to any individual

> or small number of individuals, though it may frequently do much more than repay . . . a great society (Smith, 1776/1976, IV, 9, p. 208f).

In Smith's economic model these duties confirm the authority of the state to do what the market is unable to do. Smith regarded free exchange in the market as a separate issue; it was the means of natural growth, and he contended against the state's interference in this regard. P. Sylos-Labini (1976) confirms that "in the *Wealth of Nations* competition and growth are two aspects of one and the same process" (p. 200). Moreover, he says, Smith held that the concepts of natural prices and natural rates are inseparable from the concept of competition. Smith warned against arbitrary barriers to competition. His account of mercantilism especially opposed the state's intervention in the market to restrict competition in behalf of monopolies profitable to the state.

In viewing the defense S&T market, a modern phenomenon, in the context of classical economy, one similarity is striking. State intervention has been a constant limiting condition in the market. Then as now, regulation by the state depends on the state's interpretation of how much freedom of exchange is desirable. What has changed, however, is the rationale for government intervention, and that rationale has grown to encompass a huge gray area of limiting conditions. What now drives the state to intervene in the market is a large set of social and political objectives that the market by itself cannot achieve. They include full employment, controlled inflation, the balance of payments, and a rate of economic growth that will minimize inflation; they belong to macroeconomics.

Alexander Cairncross (1976) explains the changing emphasis. We live in a world of increasing centralization, marked by a profound belief in planning and control. He concedes that the drive to control is prompted by honest motives for achieving social goals. But bending the market to political demands reinforces the assumption that it is the state, not the market that is seen as delivering the goods.

> This of itself politicizes economic decisions of all kinds and puts pressure on the state to expand its functions, its powers, and its machinery of control without much regard to the cumulative effect of the complex machinery so created (p. 128f).

Adding to this politicized intervention, moreover, is the fact that the state is also a *participant* in the market. "Nobody would seriously propose that the state should refrain from all intervention in the market . . . when the state already *is* the market for a quarter or more of national output" (p. 115).

Here is the critical feature of the limiting conditions imposed by the federal government on the market. For a substantial portion of the national economy, including defense technology, the government *is* the market. And this limiting condition significantly affects the production of knowledge.

CONSTRAINTS ON THE DEFENSE S&T MARKET

DoD spends more than half of the federal government's sponsorship of R&D, and the manner of its procurement is meticulously regulated. The regulations are ubiquitous. Their sources are described by Joseph Schmoll (1993) in a study guide for military officers published by the Defense Systems Management College. The regulations are embedded in separate acts of Congress authorizing DoD to do business, beginning with the Armed Services Procurement Act (1947) and occurring with increasing frequency thereafter.[1] These statutory regulations are implemented through executive orders and directives that carry out the congressional authorizations. Still more regulations are put into law from time to time in the form of statutory amendments to the annual legislation that appropriates funds for defense.

Within the body of law, a series of executive orders and directives, periodically updated, shape the way in which DoD does business. Chief among these are two documents that determine policy and procedures for all federal departments and agencies. One is Circular A-109 of the Office of Management and Budget (OMB); the other is the Federal Acquisition Regulation (FAR). Both documents bear on the federal government's interaction with the market. The ruling policy statement, OMB Circular A-109, establishes the concept of a mission, that is, an organizational purpose or goal, and it requires all departments and agencies to specify their needs and objectives in mission terms. Circular A-109 also levies other requirements on all federal departments and agencies: that they designate focal points for their acquisition activities, that they emphasize competitive exploration of alternative system-design concepts, and that they communicate frequently with Congress.

---

[1]The principal acts are the Armed Services Procurement Act (1947), the Small Business Act (1963), the Office of Federal Policy Act (1983), the Competition in Contracting Act (1984), the Department of Defense Procurement Reform Act (1985), and the Department of Defense and Reorganization Act (1986), referred to as the Goldwater-Nichols Act.

Given DoD's unique role as the consumer of what it buys, including R&D, these policy requirements largely determine how that role is played out in the marketplace. The mandate to establish a focal point for R&D procurement, for example, homogenizes DoD's transactions with the suppliers of what it purchases. Moreover, because of the jealously guarded turf in the three military departments, other informal restrictions are levied on the suppliers of R&D. A vendor dealing with any one military department, for example, is not always likely to be favored by the others.

The formal mandate to stay in touch with Congress actually encourages another role for DoD, since Congress furnishes the money that DoD spends. In this circumstance, DoD is a vendor competing in the market for appropriations. So the military services take seriously the instruction in Circular A-109 that mission activities communicate frequently with Congress. It is no coincidence that the congressional liaison offices of the military services are located on the ground floor of the Rayburn House Office Building, across the street from the Capitol. DoD is both customer and vendor in the national marketplace. It is the Walt Whitman of federal agencies. "Do I contradict myself?" Whitman (1855/1950) asks in *Song of Myself*. "Very well then I contradict myself, (I am large, I contain multitudes.)" (p. 74).

The rules and procedures for implementing all the policy requirements are encyclopedic. They are contained in the FAR, which consolidates the major constraints on procurement procedures for all federal departments, agencies, and activities. The FAR mainly pertains to conditions for competition among suppliers and to procedures for contract awards. This exhaustive regulatory tool was authorized by executive order in 1982, codified two years later, and then computerized for continuous updating. Moreover, the FAR is augmented in DoD's own procedural supplement (DFARS) and still further localized in supplements issued by the military services. Virtually all of these restrictions on competition and contract awards reflect prior constraints levied by Congress. To all intents, therefore, DoD is a regulator of the market in which it participates.

Perceiving these layers of authorizations and regulations, as in a carefully stratified roadbed, one senses the rigidity of the whole structure. But, like the roadbed, the structure stays whole because it can move as the earth beneath it moves; the structure has "give." Many of the authorizations and regulations are subject to interpretation. So there is room for negotiation, for shifting, for ambiguity. The flexibility lies in the interaction between the military and civilian organizations committed to national security. They are all committed, however, by way of their own objectives and agendas: advisory offices in the executive branch, the relevant committees and subcommittees of Congress, vendors (in both the public and private sectors), political lobbies, the press, and sometimes the public. All of these constituencies have a stake in shaping national security to the way they see it. Because the organizations all have their own agendas, trade-offs are inevitable.

Schmoll's guide to defense acquisition describes some of these disparate constituencies. In the executive branch, the Department of Defense, the Department of State, the National Security Council, and the Office of Management and Budget all serve the cause of national security from their different perspectives. These include maintaining a balanced force structure, fielding effective weapon systems, eliminating waste and abuse, and preventing undue congressional scrutiny. Similarly, the armed services committees and authorization subcommittees in each house of Congress serve the cause of national security and the elimination of waste and abuse. On the market side, defense contractors want profit, growth, and greater market share than they have. Universities and other nonprofit entities want a fair cost recovery and more laboratory equipment. And all R&D suppliers want the rights to their intellectual properties. The stocks and flows of knowledge that Machlup (1980) talks about are either stored or channeled according to these various, often conflicting, objectives.

CONGRESSIONAL INTERVENTION

Congress is the primary force intervening in the defense S&T market, so it is important to keep in mind how members of Congress view themselves. William Wells (1992) reminds us that they perceive themselves collectively as constitutionally coequal with the President. They are acutely aware of their membership in a highly select group, which separates them from all appointed officials. Yet the force of Congress is scarcely reducible to one vector. George Swetnam (1994) describes it as "a collage of 535 special situations" (p. 40ff). It is the essence of distributed vision. Swetnam, an engineer, spent a year with the Congress on an IEEE Congressional Science Fellowship, serving as a legislative assistant. He observes that all members of Congress have their own styles, priorities, and organizations, but that they have one thing in common. As Swetnam was formally told at the outset of his duties, they are all representatives. This common denominator accounts for the congressional priority of jobs for the constituents back home. In this respect, the inertial force of the Congress is relentless, especially regarding the national defense program.

For 50 years the legitimacy of the Cold War provided collateral for Congress to structure national defense needs into a national jobs program. Although the Cold War is now history, Congress has continued to borrow against that collateral, using tax money to provide jobs for defense on a scale no longer needed. Long before the Cold War was over, during the 1970s, the Center for Defense Information (CDI) began to warn of excessive spending for defense. An educational enterprise in the nation's capital, CDI has monitored defense spending for more than 20 years and editorialized about spending excesses in its newsletter, *The Defense Monitor*, and through TV programs. Pondering "Why We Overfeed the Sacred Cow" (CDI, 1996), *The Defense Monitor* makes a startling case. Its premise (supported by statistics) is that civilian spending generates more jobs, dollar for dollar, than military spending generates. "The irony is that continued spending to support military-related jobs is forcing budget cuts for superior job-creating civilian activities" (p. 4). The reasoning is logical. But CDI is faced with a formidable demographic fact that encourages illogical defense spending.

The *Defense Almanac* (American Forces Information Service, 1994) confirms the demography of congressional largesse. Its listing of "The Defense Presence by State" includes all 51 states, and the locations of military installations across the nation closely corresponds to the relative sizes of state populations. In addition to distributing defense dollars by state or district, the 535 congressional members share some other concerns about national defense. They care about balancing national defense needs with social needs, and about reducing the national deficit. They also care about controlling the profits of defense contractors and maximizing market competition—everywhere except in the home district. These concerns can be expressed by public laws. They can also be expressed through congressional pressures on regulatory agencies and administrative bodies that oversee defense activities. Congressional micromanagement is commonplace.

Less obviously, but with more immediate effect, congressional priorities can be directly conveyed by appropriations committees to the mission agencies that receive the appropriated money. Directives, hints, and opinions about how to spend the money are written by congressional staffs in documents that accompany the annual defense appropriations bills. Daniel Greenberg (1994) describes the genre. Known as "report language," these guidelines constitute a kind of fugitive genre of soft imperatives. Although they do not have the force of law, they carry a great deal of clout, since those same agencies will be back next year for more money.

On the subject of congressional advice, speculations abound. Old timers speak of the informal conversation as a convenient institutional arrangement. If a DoD acquisition manager and a congressional staffer were to happen into conversation about the outlines of a proposed R&D initiative, two tangible outcomes might result. The initiative might be incorporated in the appropriation of funds for some other R&D program. And a contract for some part of that initiative might ensue with a local R&D supplier in some designated congressional district. The conversation need never include any mention of a *quid pro quo*. It is simply that both parties have discovered a mutual interest in moving R&D programs forward. The acquisition manager needs dollars to advance a program; the staffer's employer needs to channel dollars into the home district. The acquisition manager might even help the staffer draft the legislation. Both parties understand the culture.

The mutual understanding of the program manager and the congressional staffer is the point of interface between two large institutions. The defense R&D enterprise is not singular in this regard. The scenario is pervasive in federal transactions for R&D, although it has different versions, depending on who is pressing for the program. One version of the scenario is described by Arthur Squires (1986), who has studied the federal R&D system from inside its culture. In his version, a federal agency has become the province of a small set of legislators, so that everyone outside their districts is disadvantaged. "Some bureaucrats are frank about this limitation," Squires says. "An academic who had submitted a proposal to one agency of this type heard the advice: 'We like your proposal. Now you must find yourself a Congressman.'" (p. 131).

The arrangement of a *quo* for the *quid* could, of course, be an oversimplification. Yet the primacy of jobs typically tempts Congress to outright intervention, particularly in defense technology, which is so labor intensive. One highly publicized example of this temptation occurred in the defense authorization bill for FY 1996. Notwithstanding its commitment to the military drawdown, Congress extended funding for the B-2 stealth bomber. Both DoD and the Air Force had expressly testified that no more B-2 bombers were needed or wanted, but Congress had a different priority, bolstered by the national labor force which the plane supports. David Maraniss and Michael Weiskopf (1995) recount the history of the lobbying campaign that won the needed votes in Congress. Conducted by plane's airframe manufacturer, the campaign was supported by hundreds of subcontractors and thousands of suppliers spread throughout 48 states.

In the negotiating process for the same authorization bill, Congress revived the Strategic Defense Initiative, scarcely three years after it had been terminated. In terms of its original purpose, the projected space shield against a massive Soviet nuclear attack was a technological failure; the sustained attempt to design a system of space-based interceptors literally never got off the ground. Yet the defense authorization bills in both House and Senate for FY 1996 specified the expenditure of some $600 million above DoD's requested funding for antimissile technology. The declared intent was to resume work on exotic spaced-based lasers and other components of SDI, notwithstanding the ABM treaty, which prohibits their deployment. As reported by Weeks (1995), momentum for the new R&D was supplied by the Congressional Missile Defense Caucus, organized within the House expressly to build support for a sequel to Star Wars.

Congress has sometimes engaged in even more acute micromanagement of defense technology. Richard DeLauer (1986) recalls one publicized occasion wherein Congress preempted the entire defense acquisition process. DeLauer headed the program for weapons acquisition and R&D at the Pentagon during the early years of the Reagan administration. At issue was the development of a gun for division air defense, called DIVAD, which was actually a creature of Congress. The consensus of the congressional defense committees was that the Army had taken too long to develop a field weapon of this type. "Congress decided to make the Army use existing components to get a new air defense system deployed quickly" (p. 58f). As DeLauer explains, the Army duly cobbled the chassis of a 109 howitzer, the radar from an F-16 fighter, and an existing gun. But when developmental tests of the weapon did not meet congressional expectations, despite its promising performances, the Defense Reform Caucus in Congress killed the program.

Congressional preemption of the market occurs through another, even more deliberate, procedure called "earmarking." Acknowledged in more formal language as "congressionally directed funding," this procedure involves the attachment of riders to legislation; the riders allocate money directly to specific projects in designated congressional districts. These pork-barrel programs range over the entire economy, including federal courthouses equipped with showers in the judges' chambers (Guy Gugliatta, 1993) and a theme park for a desolated downtown area (Kent Jenkins, 1993). Since national defense needs command about half of the discretionary spending in the federal budget, earmarked funds for defense programs are commonplace. They even affect funding for basic research in universities, thereby exempting the recipients from the traditional peer review common to the science community.

Earmarking also exempts the recipients from the overall planning process in DoD. Despite DoD's priorities for key fields of research during a period of decreasing budgets, earmarked funds outside the planning process have significantly increased since 1986. In that year, Congress established the University Research Initiative (URI) to broaden DoD's research base. Then it began to engage in directed funding of the program. The Congressional Research Service has documented the extent of this trend in Michael Davey's CRS Report for Congress (1994). "Between FY 1991 and FY 1994, Congress appropriated a total of $1 billion for URI, of which $342 million was congressionally earmarked funding" (p. 14). During FY 1993, the report continues, earmarked funding accounted for 56% of the program's budget.

At issue here is not the value of programs funded in this manner. (One old timer at DoD reminded me that the USS Wasp and USS Hornet had been earlier pork-barrel projects that turned out pretty well in World War II.) At issue, rather, are two other considerations. One is the congressional reduction of the proportion of Federal funds that would otherwise be available for competition in the marketplace; the other is the congressional increase of Federal funds which it exempts from coordinated planning efforts. Earmarking is simply a more obvious form of directed discretionary spending whereby Congress intervenes in the market.

REGULATIONS, PAST AND PRESENT

As for technology programs already under development within the jurisdiction of the defense acquisition system, Schmoll's (1993) guide dramatizes the plight of R&D management. With a graphic illustration entitled "The Tortured Triangle," it depicts the impact of all the political constituencies on the DoD program manager (p. 4). This vivid display is a maze of vectors. At the center of the triangle, the program manager (PM) is impaled on three sets of arrows that travel to and from each corner of the figure, labeled Congress, Executive, and Industry. Twelve sets of two-way arrows, connecting each point to every other, are labeled with different regulations or organizational procedures. The surrounding space is filled with bursting star shells that represent other special-interest groups. The maze of vectors is a remarkable piece of candor (see Fig. 5.1).

The defense S&T market is so beset by regulations as to make it unique among markets. This peculiar condition is a Cold War phenomenon. It began at the close of World War II, during which the national productivity had been totally mobilized for defense. Sapolsky (1972) describes what happened. "This war, unlike others, was not followed by a period of near

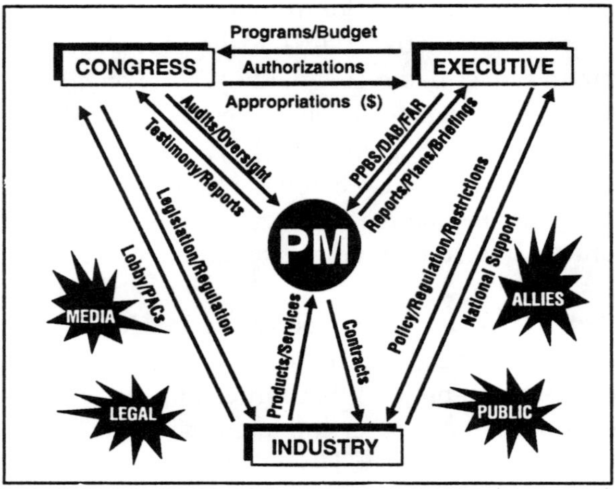

FIG. 5.1. The tortured triangle.

total disarmament. Instead, it was followed by a period of sustained arms purchases. Firms mobilized or created during the conflict had the opportunity to continue weapon developments with the expectation of profits on future production contracts" (p. 77).

The stakes were high, and the possibility of monopolies restraining trade were great, so federal regulations were necessary. Congress became the market's regulator, but DoD was the consumer of the market's products. The dilemma over priorities and jurisdictions was immediate, and it has persisted. Committees in Congress and regulatory boards in the executive branch duly strive to regulate the market in which DoD participates as a customer.

In assessing the magnitude of these layered authorizations and restraints, one might ask how far government regulation has penetrated the market since Adam Smith took note of it. As to the fact of blatant intervention, there is a startling similarity between past and present. Smith cited—and indicted—the encyclopedic strictures of the British crown on the wool trade of Kent and Sussex counties (1776/1976, IV, 8, 167f). Beginning with the shearing of the sheep, those regulations categorically tracked and limited every single exchange of fleeces within 15 miles of the sea. The state's disposition to control the market has not changed very much in the past two centuries. But what has changed is the quantum increase in the number of constituencies that have become involved in the state's regulation of the market.

Because of these plural forces, governmental regulation of the market for national defense has grown immensely complex since Adam Smith charged the state with protecting its people from invasion. Although a line might be traced, delineating the respective spheres of the modern defense market and the state, the two would blur hopelessly. In the transfer of goods for money, an older concept has been modified by a newer one; the older concept of the free exchange of values determined by market forces has been modified by the newer concept of an authorized exchange of allocated values. The result of the change is a contrived market. Such is the market that both enables and circumscribes the Department of Defense. It is a byzantine construct that reflects the contrivances of many parties within the public sector. The result is a set of constraints that bear only faint resemblance to the classical economic model of a free market.

What has gradually evolved from this governmental process is an R&D market freighted by limiting conditions that reflect social needs and political objectives. Such needs and objectives extend well beyond the issue of national defense. The most recent of these macroeconomic directives pertain to the mandate for defense conversion. They are intended to create new channels for the acquisition of defense technologies through protocols designed to make the national economy more competitive. But these newest directives have become highly politicized, and their outcome is yet to be assessed.

Recalling his year spent living on the brink as a legislative assistant in Congress, Swetnam said that it made him understand how nearly ungovernable the United States really is. Perhaps the key word is *nearly*. From an economic perspective, what is remarkable, amidst all of these allocated values, is the market's resilience. "The market endures change by exploiting it," Cairncross (1976) has observed. "It is a great virtue of the market that it provides a testing ground for responses to new influences, for eliciting and trying out new alternatives and demonstrating which new ideas have greatest survival" (p. 133).

6

The Color of the Money

The distribution of knowledge generates yet more knowledge, as Machlup (1980) has indicated. Flows of knowledge from one party to another increase stocks of knowledge for both parties. Flow is a critical concept, and it can be tracked, as Machlup says, by "the measuring rod of money" (p. 175). To the extent that R&D converts information to knowledge, Machlup's rule of knowledge production attains at the Pentagon. The flows of knowledge follow the flows of money.

The magnitude of funding for defense R&D is fairly evident from the statistics gathered for NSF and cited in chapter 1. Since 1955, the money appropriated for defense R&D has averaged over 61% of the total federal R&D expenditure. From a different perspective, the figures are confirmed by Richard Rowberg (1995), reporting for the Congressional Research Service. Although the cost in current dollars changes annually, the average proportion of funding has remained stable. The appropriation for defense R&D in current dollars for FY 1996 was about $38 billion.

But the annual subtotal of defense R&D money expended on research is a more elusive sum; as deduced from several different budgets within DoD, it accounts for about $8 billion, or about 20% of DoD's comprehensive R&D budget. This chapter tracks the flow of those research dollars through the bureaucracy that turns them into knowledge.

One old-timer wryly observes that the chief functions of this bureaucracy are to keep the dollars rolling in and to obligate them before the annual deadline. In this caricature, the bureaucracy is a simple input–output device with two functions: getting and spending. The procedure of getting can be simply described; it recurs with each annual federal budget. After two years of preparation and negotiation among the military services, the DoD, and the congressional committees for authorization and appropriation, the dollars are committed by Congress. The money appropriated for R&D flows to five or six designated agencies within DoD and then to those activities that both perform R&D and engage R&D vendors.

But the subject of this chapter is spending, particularly the spending for research and its consequence for knowledge production. Research is a domain within defense R&D, but the amount of money spent on it depends on how one defines it. The size of the research domain is determined only partly by the annual R&D budget, so defining it in terms of dollars is merely a gross beginning; the task of definition entails more than arithmetic. The nature and scope of defense research are determined through the interaction of different planning authorities. These are chiefly the Undersecretary for Acquisition and Technology and the Office of the Director of Research and Engineering (DDR&E); they also include R&D planning offices within each of the military departments and the mission activities in the field. As a result of their various purposes, the domain of defense research has been shifting during the past 40 years, and it continues to do so. Defining defense research, therefore, is rather more like an exercise in calculus than in arithmetic.

The task of definition entails several tours through the same territory, each one providing a different perspective. An analogy comes to mind. Taking different tours of the palaces and churches and museums of an old city helps the serious visitor to gain a cumulative understanding of the culture. Similarly, we will embark on several brief tours through the defense R&D establishment. Each one focuses differently on the flow of money to the R&D activities. The first tour views the kinds of research that are funded through DoD's annual budget. The second tour takes a different route; it describes an evolving domain of research that led to the present Science and Technology (S&T) Program. The third tour centers on organizational interactivity; it highlights the kinds of alignments that result in the transfer of information. Together, the three tours reveal a systemic process that involves the disbursement of money, the dissemination of information, and the conversion of information to knowledge.

TOUR 1: THE ALLOCATION OF R&D MONEY

The money flow is evident in the defense R&D budget structure, which channels funds to the different research and development activities. R&D represents one of the 10 program elements (PEs) in DoD's overall budget classification; the category label for R&D is PE 6. In 1979 the Office of the Secretary of Defense (OSD) regrouped some earlier classifications of R&D funding into a numbering system that included six categories of expenditure, numbered 6.1 through 6.6. This set of categories delineated a range of R&D efforts that still exists:

6.1 Research
6.2 Exploratory development
6.3 Advanced development
6.4 Engineering development
6.5 Management and support
6.6 Operational system development

Not surprisingly, the military services all interpreted the categories differently, and over time the differences caused enough confusion to warrant piecemeal changes in the scope of the categories. As a result their designating numbers have been moved around somewhat. The most recent revision in the numbering system was specified in DoD's budget for FY 1995, but the categories have remained essentially the same. For our purpose, the historical numbering offers a more accurate view of the problematic boundaries of research that have characterized defense R&D.

By designating the current money available to each category, the numbers tag the dollars that flow into the entire R&D network. The tags are so familiar to members of the R&D community that they commonly use the numbers alone to designate both the categories of money and the types of research or development that the money funds. The numbers provide an operative metaphor with instant value. When someone asks, "What is the color of the money?" the answer is 6.1, 6.3, or whatever it happens to be.

The budget classifications were originally defined in the DoD Instructions, *DODINST 7720.16* (DDR&E, 1979). They are not pristine. The definitions of each category vary in density. They are elliptical; they occasionally overlap, and they are full of qualifications, which allow for maneuvering room. But the definitions do convey an obvious rationale, which is to represent the whole range of R&D as an incremental process. The one exception to the incrementation is item 6.5 (management and support), the money in which is spread around to support the other five items. Because it is different in kind, it appears to be out of place in the numbering sequence.

One noteworthy characteristic of the official definitions is the lack of any reference to the customary polarity of basic research and applied research; on the contrary, they even muddy the distinction. They posit increments of inquiry all the way from basic research to the engineering development of weapons systems, even after the funding for those systems has been authorized. The designating numbers posit a continuity of inquiry that stretches from concept to product. Excerpts from the *DODINST* confirm this incremental notion, which is deep in the folklore of defense R&D.

In the list that follows, the quoted language is excerpted from *DODINST 7720.16* (DDR&E, 1979) Part V, Budget and Accounting Classifications (Attachment C, p. 516f). It is what I call a tier 2 document; addressed to insiders, it provides only a minimal context. My annotations follow each excerpt.

6.1 Research

> " . . . provides fundamental knowledge" and "forms parts of a base for . . . subsequent exploratory and advanced developments"

Research at this stage is curiosity driven. It entails looking at a natural phenomenon and trying to see how it works. For example, 6.1 research might entail growing a new semiconductor material. That might be the end of it. On the other hand, a semiconductor of that new material might later be used or tested in some electronic device.

6.2 Exploratory Development

> " . . . may vary from fairly fundamental applied research to quite sophisticated bread-board hardware, study, and planning efforts." It is "pointed toward specific military problem areas" with an emphasis on "developing and evaluating the feasibility and practicability of proposed solutions and determining their parameters."

This is the first stage of advancing an understanding of a natural phenomenon to some application. Designing a microwave amplifier might fit this stage of R&D. But the effort would not be obligated to serve any particular engineering system.

6.3 Advanced Development

> "The prime result of this type of effort is proof of design concept rather than the development of hardware for service use. Projects in this category have a potential military application."

There is a range of possibilities here. Designing a radar receiver might qualify for 6.3 money; so might building a radar breadboard. Whatever the artifact, proof of concept might involve some form of demonstration, and it would entail the preparation of a data package to help evaluate the risk of proceeding to full-scale development.

6.4 Engineering {And Manufacturing} Development

> "... projects in full-scale engineering development for Service use but which have not yet received approval for production ...."

The output of this stage of R&D would be the prototype of a device or a system, a radar receiver, for example. It would be expected to include a validated documentation package to support the production of that device or system.

6.5 Management and Support

> "... support of installations or operations required for general research and development use [including] test ranges, military construction, maintenance support of laboratories, operations and maintenance of test aircraft and studies and analyses in support of the R&D program."

This sprawling category of funding is not sequential; it pertains to the R&D efforts categorized in the other items. It supports test ranges and evaluating facilities stationed around the world. It also supports the DoD's R&D library at the Defense Technical Information Center (DTIC).

6.6 Operational System Development

> "... projects in full-scale engineering development ... which have received approval for production ...."

This final stage of the R&D series looks much like the color of 6.4 money. The difference is that the device or system in this case has already been funded for production. Once a certain radar receiver has been approved for production, for example, how should it be manufactured? What is the final engineering plan? Money in this category funds efforts to answer such questions.

These categories represent one dimension of the R&D budget. But there is another. Since the acquisition process is goal oriented, the DODINST specifies discrete kinds of effort to which the tagged money may be allocated. They are related to any of six different, but concurrent, military objectives:

1. Technology Base
2. Advanced Technology Development
3. Strategic Programs
4. Tactical Programs
5. Intelligence and Communications
6. Defense-wide Mission Support

These discrete objectives constitute a broad range of allocations, collectively called Research, Development, Testing, and Evaluation (RDT&E). Acknowledging this second dimension of the budget, it is common usage in the R&D community to compound the basic question: What is the color of the money, and what is the mission? The budget construct can be represented as a matrix that combines the six R&D categories (6.1–6.6) with the six RDT&E categories (see Fig. 6.1). As with any matrix, this construct

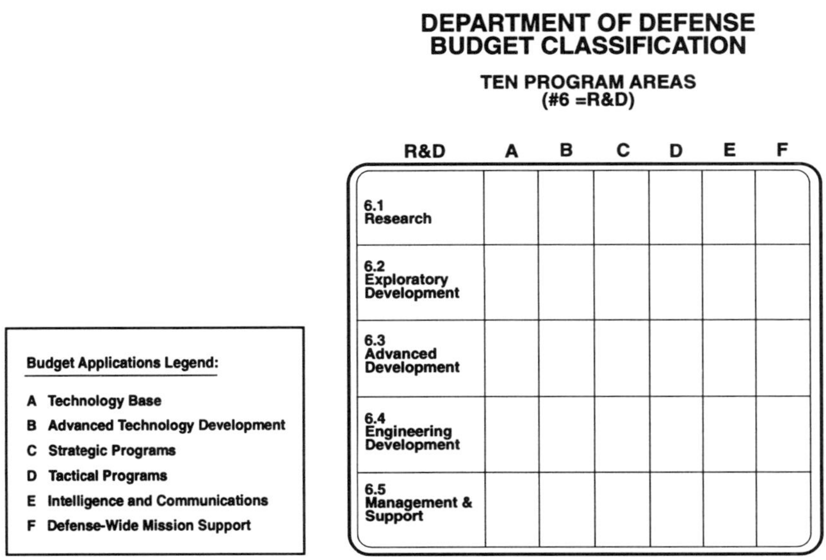

FIG. 6.1. Defense R&D budget categories.

juxtaposes two kinds of value; the six money categories are different in kind from the six RTD&E categories.

The budget matrix can be likened to a bus schedule, which plots departure against arrival. Each box down the left column represents a bus, and each box across the top tells where it is expected to arrive. In the budget matrix, each box down the left column represents a departing category of money, and each box across the top pertains to its place of arrival. But the similarity stops there; the dissimilarity is more telling. The bus that is referenced in each box of the bus schedule is discrete, and its arrival at each destination is unique. In the budget matrix, however, the money in one tagged category can be allocated to more than one kind of mission activity. Moreover, as defined in *DODINST*, the mission categories are not exactly parallel; they overlap in places, allowing for discretionary decisions by a good many bus drivers along the way.

There are only two instances in the R&D budget matrix where the money in a specified category arrives at a single, designated place on the money schedule. These invariable assignments of money pertain to the first two R&D categories: Research (6.1) and Exploratory Development (6.2). These two constitute DoD's technology base, a knowledge reserve that continually accretes and that serves possible future applications (see Fig. 6.2).

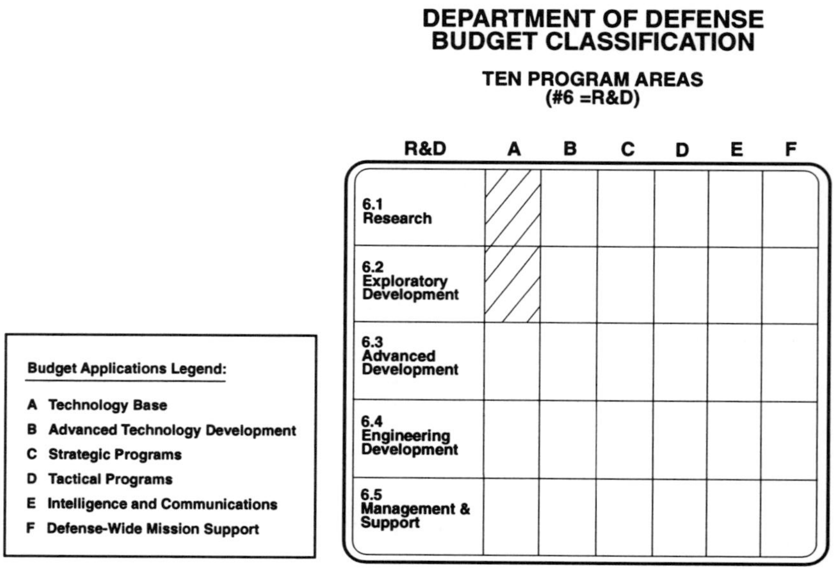

FIG. 6.2. Department of Defense technology base.

A note about the larger context of DoD's technology base is appropriate. Although huge, it is only part of what is referred to as "the defense technology base." They are two different orders of magnitude. The distinction is explained in *The Defense Technology Base* (OTA, 1988).

> The *defense technology base* is an accumulation of knowledge, skills, capabilities, and facilities, while the *Defense Department's technology base* programs are collections of thousands of individual research projects funded through the DoD budget, the results of which contribute to the defense technology base (p. 7, italics added).

The distinction is historically important. For many years, the separation of the two technology bases was accepted in matters of technology policy. The desirability of their merger is now the rationale for the concept of *defense conversion*. This central issue is examined in OTA's follow-up report, *Holding the Edge: Maintaining the Defense Technology Base* (OTA, 1989).

TOUR 2: THE CHANGING DOMAINS OF RESEARCH

The budget matrix is a static construct, one that conveys an illusive sense of order. This next tour examines what happens to the budgeted money that was packed in those neat boxes. Disorder sets in immediately, because one box of money can be sent to different destinations; it can also be redistributed as soon as it arrives—and sometimes before that. Tour 2 requires a seat belt.

The nature of R&D management changed substantially after World War II. The integration of discrete technologies into systems and subsystems necessitated networks of management organizations within DoD to oversee the process. And as funding escalated in response to Cold-War developments, so did the regulatory constraints. Over the years, the expansion of management organizations and the increase of limiting regulations have enormously complicated the task of coordinating military acquisition activities. Management has had to engage in nearly continuous articulation: repositioning component R&D activities and explaining the new alignments. This strenuous exercise has brought the defense acquisition system to a state of exquisite intricacy. "The process is in terrible shape," John Deutsch (1993) declared, soon after becoming Under Secretary of Defense for Acquisition and Technology.

It's not the result of past people having been either stupid or dishonest. It's the result of an infinite number of regulations, different, well-intentioned constraints imposed on the system. It's become so bound up in its own process that it can't get anything done efficiently . . . ." (Deutsch, 1993, p. 8).

As part of the defense acquisition system, R&D management has undergone these same stressful conditions. Moreover, as the technology of integrating weapons development has become more complicated, so has the task of managing it. As a result of these developments, R&D management has necessarily engaged in redefining itself and reconfirming its component activities. One particular consequence of such reorganization, since 1980, has been the continual reshaping and resizing of the domain of research. As might be expected, the reorganization process has mainly taken the form of redefining jurisdictions in the management of R&D money.

These jurisdictions furnish the reference we need to track the changing domains of defense research. In the introduction to this study, the jurisdictions are represented by a static figure of contiguous ellipses that broadly categorize different knowledge-producing activities. Repeated here from a somewhat different perspective, that figure provides a convenient reference for delineating the domain of research (see Fig. 6.3).

Our present perspective concerns change through reorganization over a period of time. The changing size and shape of the research domain—a component of R&D—can be tracked through the reorganizations of R&D management.

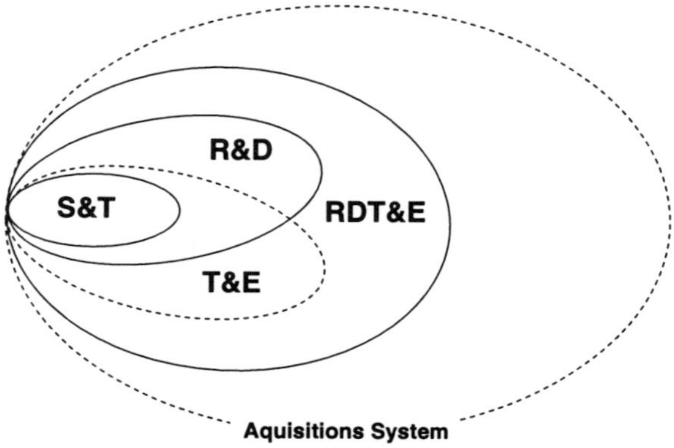

FIG. 6.3. Knowledge-producing activities.

In the beginning, at DoD, the fiscal management of technology development was centralized. During the first 10 years after the National Defense Act of 1947, the Secretary of Defense directly allocated research funds. Then, after the launching of the Soviet Sputnik in 1957, a sudden emphasis on R&D prompted the creation of the Office of the Director of Research and Engineering (DDR&E) at the undersecretary level. Jurisdiction over all research was passed to that office in 1960, and it remained there for 20 years. During that period the domain of research consisted of R&D categories 6.1 and 6.2. Results of research funded in those categories yielded DoD's technology base, the growing reserve fund of knowledge on which later R&D activities might draw.

Beginning in 1980, as part of a series of reorganizations at the Pentagon, the defense bureaucracy expanded. Although DDR&E was moved successively downward in the management hierarchy, it maintained direction of the research domain that was actually growing. The domain was redefined to include R&D category 6.3, Advanced Development. This was a huge enlargement. Then it was narrowed to include only part of 6.3. The net effect of these organizational changes was to divide category 6.3 (Advanced Development) into two, namely, 6.3A and 6.3B, and to draw a significant jurisdictional boundary between them. Category 6.3A became Advanced *Technology* Development; 6.3B became Advanced *Systems* Development. With this change, the earlier domain of research was expanded to include R&D categories 6.1 through 6.3A. This organizational change shaped the new domain of S&T.

The change in the grouping of budget categories over a 40-year period reflects revisions in the concept of what constitutes research in an R&D environment. It is a subject that has recurrently engaged budget analysts, congressional committee staffs, and public policy analysts—all from different perspectives and for different purposes. The new domain of S&T is a pragmatic response to policy questions about the organizing of technology. After the Cold War, S&T became the focus for the objective of making technology a force multiplier, as we have seen in *Defense Science and Technology Strategy* (1994). The S&T concept has already acquired a new public presence in having become the S&T Program for achieving it.

No matter what its public presence, however, S&T has an institutional identity. Within the range of R&D activities, S&T is the largest domain of generic research, as distinct from research that pertains to definable weapons systems. Although it has not been categorized as a single budget entity, it has represented annual expenditures of about $8 billion a year since the 1980s. But its budget is not what distinguishes the S&T domain. Rather, its distinction lies in its independence of all funding for weapons programs. Its coherence lies in its relative autonomy with respect to such programs rather than in the activities that it supports. This distinction has important implications in the R&D infrastructure: the domain of S&T represents generic research.

Although the three funding categories that constitute S&T appear to represent stages of research, the activities they support are not serial, and they do not necessarily relate to each other. They even serve different objectives, which pertain chiefly to their military implications. As distinct from research in the other two categories, for example, 6.1 research may or may not be relevant to some distant military application. The distinction between the other two is that 6.2 research is focused on "key technologies," and 6.3A research supports what are called "major thrusts." Although both have military implications, there is a difference. Key technologies are those seen at any time to be priorities in the possible development of future weapons systems. They are DoD's canon of critical technologies. But without reference to particular technologies, "major thrusts" represent possible military *capabilities*, such as conducting global surveillance, controlling the use of space, or achieving and conveying near real-time knowledge of military events.

Another kind of distinction also marks the three categories of S&T; it pertains to relative risk in research. Expressed as a percentage, risk is customarily an inverse estimate of the probability of positive outcome; the estimate is based on the time and cost entailed in producing that outcome. A 90% risk, for example, could indicate a low possibility of positive outcome, given the time and cost. Or it could indicate a long lead time and therefore a higher cost no matter what the outcome. But the three funding categories (6.1, 6.2, and 6.3A) are not subject to risk estimates. Since they are not grounded to any particular technologies, their sequential numbering does not correspond to any systematic estimates of relative risk.

The risk factor can vary widely within any one of these three categories. In the domain of S&T, risk depends not on the stage of research activity as defined by funding, but on the discipline within which the research takes place. Projects in optoelectronics, for example, are likely to yield positive results much sooner than do projects in advanced materials. Research on a laser application in optical computing might pay off within a year. But a project in the application of the elastic properties of superplastic steel might take several dozen years, possibly without any payoff at all.

In any case, risk—the uncertainty of outcome—is a real factor in the way money changes hands after it has been allocated. It affects the flow of money—and information—among the R&D activities. The interplay of ideas among the activities is a response to risk; it concerns the budgetary distinction between money for generic research and money for systems. That distinction is definitive in the management of defense R&D, for it has engendered two separate intramural communities. One is the generic research community, and the other is the weapon-program community. Although command structures in the three military services differently separate the two communities, the distinction between them remains. Because risk is a factor at every stage of R&D, its presence generates lateral flows of information between various generic research groups and various weapon-program groups. The color of the money is critical.

The distinction between 6.3A money and 6.3B money is what has separated the two communities over the years. It marks the far boundary of generic research and the near boundary of systems development, and that boundary is porous. As in most boundary conditions, the system operates erratically. Although intended to separate one kind of effort from another, the distinction between generic money and systems money sometimes has the opposite effect. It can actually result in interaction among the different research management communities. As in most bureaucracies, organizational jurisdictions follow the money, and, in this case, once the money starts flowing into the system, its various colors tend to merge.

Here is how that happens. With one notable exception, which we will turn to in a moment, generic research money to be expended within DoD is allocated primarily to the directors of defense laboratories. The money for systems technology is allocated primarily to program managers of weapons systems that are under development. Because an uncertain outcome gets more expensive once a weapon system is under development, avoiding risk becomes critical for the systems program manager. Typically, an unforeseen engineering problem occurs, provoking the need for additional research, but generic research money is not in the budget for systems. The program manager must therefore find ways to contract or barter services with laboratory managers to accomplish the needed research.

This kind of lateral transaction subverts the neat, discrete boxes in the budget matrix. Repeated often enough, it makes the system messy. But it gets the job done. Such transactions, moreover, distribute knowledge among researchers with different missions, increasing the stocks of knowledge all around. For operational reasons, therefore, the apparent linearity of the budget categories 6.1—6.6 turns out to be illusive. The orderly allocation of funds is one matter; the *ad hoc* bartering in their disbursement is another.

Disbursement accounts for the dissemination of information and its conversion to knowledge. This process occurs prior to the existence of a weapon system and throughout its life cycle as well.

TOUR 3: THE INTRAMURAL CUSTOMERS AND VENDORS

Within DoD's organizational hierarchy, six agencies oversee most of the in-house RDT&E activities. These include the three military departments (which are regarded as agencies), ARPA, the Defense Logistics Agency, and the Corps of Engineers. Most of the 90 or so R&D activities report to the military departments, and many of them commonly interact with each other. In the role of either customer or vendor, they exchange information and collaborate in converting it to knowledge.

Like the previous tours of the defense R&D establishment, this tour of the in-house activities begins with the budget. Money allocated to R&D flows unequally into two pots: the lesser share going to intramural activities, for R&D performed in-house, and the greater share going to extramural vendors beyond DoD. But the in-house activities do not retain all of their lesser share. Some of it stays in-house, and some of it passes through, finding its way into various external pots. The effect of the flow is to increase the external share of the money to about 75% of the original budget allocation. These are the main channels of distribution through which money flows. But they twist through many rapids along the way. As the intramural money passes downstream, the budget categories are no longer reliable indicators of where it goes.

As for quantifying the annual money flow, actual dollar amounts are duly recorded several years after being budgeted; expenditures are reported to OMB, where they are compiled and forwarded to NSF for occasional publication. The whole procedure takes time, and even then the money spent for generic research cannot be accurately known. As reported by Meeks (1994), Defense R&D expenditures for FY 1993 totalled $37.9 billion. Of this amount, $11.1 billion (about 30%) was charged to budget categories 6.1, 6.2, and 6.3. But because of the reporting categories, the money that was actually spent on generic research (6.1, 6.2, and 6.3A) can only be estimated. If we assume that the money for Advanced Development (6.3) was evenly divided between generic research and systems development, we have a basis for estimating. In this case, the dollar total for generic research would have been about $8 billion, or about 20% of the total defense R&D expenditures.

Of this dollar amount for generic research, $1.3 billion went into basic research during FY 1993. Mostly through grants, it was distributed by the Army Research Office (ARO), the Office of Naval Research (ONR), and the Air Force Office of Scientific Research (AFOSR). All three research offices oversee programs in mathematics, engineering, and the sciences, which they coordinate with mission priorities. Differently situated in the command structures of the respective military services, they have somewhat different organizational responsibilities. But they all court the scientific community, chiefly in academic institutions, and they periodically update complete descriptions of their programs for prospective grantees.[1]

The bulk of the money for generic research during FY 1993 was spent on Exploratory Development (6.2 money) and Advanced Development (6.3A money). It went to a variety of R&D laboratories and T&E centers, which vary widely in staff size and budget allocations. Among these activities, the larger laboratories perform dual roles in the R&D market; they buy and sell research. As customers, they buy research from other DoD activities, from universities, and from industry, and as research vendors, they sell their own services to some of the same parties. Herein lies much of the complexity in the flow of information.

Some of the laboratories are deliberately underfunded; they must seek support beyond the budget process, much in the manner of university science and engineering departments. The Naval Research Laboratory (NRL), the Navy's largest, operates in this way; it transacts with other institutions in both the sale and purchase of research at the same time. Similarly underfunded, the larger Army laboratories sell their research as well. For many years, for example, the Harry Diamond Laboratory (HDL) was partly funded by the Army for research in the whole range of categories 6.2 through 6.5. But it also sold research services to other DoD activities and to private industry. In the Air Force, the Wright Research and Development Center (WRDC) has received external funding from NASA and DOE.

The deliberate underfunding of RDT&E activities has had fortunate consequences for knowledge production. It has forced laboratory directors

[1]All three offices update their program announcements from time to time. There are current editions of *Research and Development Opportunities*, U.S. Army Laboratory Command; *Office of Naval Research Guide to Programs;* and *Research Interests of the Air Force Office of Scientific Research.* A brief description of opportunities in DoD's overall research program is *Department of Defense Research Program and Related Opportunities in Science and Engineering Education.* The brochure is occasionally updated.

to seek and exploit new channels of funding, and the flows of knowledge have consequently multiplied. Underfunding has put the laboratories in competition with one another as they seek to sell their research services. As a result, they necessarily compete with other research vendors, including universities and commercial firms. The competitive environment has also engendered modes of collaboration among the RDT&E activities, involving pooled dollar resources and personnel exchanges for specific projects. These bilateral arrangements accelerate the exchange of information and the interpretation of what is exchanged.

I mentioned earlier an exception to the policy of allocating generic-research money directly to the R&D laboratories. That exception is the Advanced Research Projects Agency (ARPA), a major catalyst for knowledge production in the defense R&D community. Created by Congress in 1958, in response to the Soviet launching of Sputnik, ARPA was chartered to be DoD's central R&D organization. The agency was President Eisenhower's vision. He wanted a different type of organization to generate revolutionary technology, one that would cross disciplinary lines and transcend organizational constraints. Reporting directly to DDR&E, ARPA has remained independent of the military departments.

The primary responsibility of the agency was to seed and nurture defense technologies for the future. It has continued to do so despite changing political environments. To reflect its emphasis on military technology during the 1970s, the agency's name was changed to include "defense," and ARPA became DARPA. Later, in 1993, to reflect DoD's new emphasis on dual technologies after the Cold War ended, DARPA became ARPA again. But names notwithstanding, the agency has remained steadfast to its mission for defense. Insiders know that "ARPA" is really pronounced DARPA with a silent D.

Language describing the agency has a familiar ring. Its mission has been to pursue imaginative and innovative research and development projects offering significant military utility, with emphasis on scientific breakthroughs and revolutionary approaches. What is notable about these phrases is the authority that DARPA conveys within DoD. Victor Reis (1991a), DARPA's director at the time, used them to describe this one agency to a House subcommittee on R&D in the armed services. And just three years later they were prominently echoed in *Defense Science and Technology Strategy* (1994) to describe the thrust of DoD's *entire* S&T Program.

ARPA is agile. It is sometimes described as a small group of people held together by a travel agent. The core of the organization is a staff of some 80 program managers who are scientists and engineers. This group directs teams of experts from both public and private sectors. Reis (1991b) describes the extraordinary range of their responsibilities. They include planning research programs, managing program funding activities and competitive contracting efforts, and soliciting ideas from universities and commercial firms. They also include the marketing of new program ideas to DoD's top management. As subject specialists, they are also expected to stay abreast of the latest developments in their fields. In the defense R&D community, these resident experts are legendary.

ARPA's focus on scientific breakthroughs is obviously reflected in DoD's canon of critical technologies. In all aspects, ARPA's program is inherently interactive, engaging other R&D entities all along the way. The Desert Storm operation was a definitive harvest of the ideas that DARPA had sown and nurtured. Reis (1991b) cites the large, ground-based, phased-array radars; stealth aircraft; wide-area airborne surveillance, self-correcting planning tools for transportation and logistics, friendly identification devices, and armored sheathing of advanced composite materials to wrap around vehicles. The lead time from initial research to a deployed system varied widely, from 30 years for the phased-array radar to only a few months for the armor sheathing.

One common element underlying all these knowledge products is ARPA's policy of engagement with the three military agencies and their various R&D activities. ARPA develops ideas to a viable stage and then transfers them to the appropriate activities to oversee the procurement of R&D. It has been a durable policy; its rationale dates back to the National Defense Research Committee (NRDC), under Vannevar Bush, during World War II. NRDC took the initiative in setting research priorities and assigning resources to support them. As noted earlier, Stewart (1948) recalls that "NRDC asserted the right to exercise an independent judgment as to the projects which it should start as well as the method of attack upon them" (p. 18).

ARPA manifests a policy of technology "push," as distinct from demand "pull." It explores new ideas without committing to their full scale development, as opposed to responding to needs for improving current weapon systems. With the onset of defense conversion, ARPA's policy of technology push became more timely than ever. The defense drawdown and the prospects of leaner budgets have prompted a new acquisition strategy at DoD, emphasizing more distant technological horizons. After the Cold War, ARPA quickly became an ideal mechanism within DoD for changing an old military habit of committing resources primarily to the production and deployment of weapons. The new vision is to commit more resources to the advanced development of new ideas. "This decision to explore can be clearly separated from the decision to develop and the decision to produce or deploy." So said another DARPA director, Gary Denman (1992), in his report to a congressional committee. Therein he underlined the importance of maintaining generic research:

> DARPA's mission provides the Department of Defense with the ability to explore new concepts in a manner that clearly separates the agency from bureaucratic inertia, roles, and mission dogma, and the syndrome of "that which is possible automatically becomes that which is required" (p. 2).

Denman's statement highlights ARPA's policy of maintaining a critical independence for R&D "with relatively modest investment where leverage is greatest—in the conceptual stage." This powerful concept continues to guide knowledge production in the defense R&D establishment.

7

Customers and Vendors: Dyads in a Dance

So far, we have tracked the flow of money budgeted for generic research to DoD's in-house R&D activities. Some of it is retained for R&D services performed in-house and for overhead costs. But most of it flows out to vendors in commercial firms and in universities and other nonprofit organizations. This outbound flow of money converts DoD's intramural activities into customers for R&D services. The interaction of customers and vendors in this extramural market is our present subject. The more we know about their interaction, the more we will know about knowledge production, since the one engenders the other. Customers and vendors readily exchange information, and the exchange process catalyzes their mutual interpretation of it. In so interpreting, they coconstruct knowledge.

We turn, then, to the kinds of interaction that mark the extramural R&D market, with an eye to patterns that engage these parties in knowledge production. We begin with a profile of the external vendors; it is abstracted from the DoD databases for extramural R&D contracts. We then view the kinds of limiting conditions that warp the market. Borne primarily by the vendor, they are the stringent regulations imposed on the competition for contract awards and on the services that the vendors perform. Finally, we see the kinds of relationships among customers and vendors that exist within these limiting conditions.

The observations that follow bear an implicit proposition about where and how knowledge is produced. Granted, the R&D process yields knowledge; it yields findings and also conclusions about those findings. But, prior to the actual R&D, the business of procuring it also produces knowledge. The information that is transferred in the market transactions and the knowledge gained thereby are less obvious than the findings of the research that is bought and sold. But this kinetic knowledge, if I may call it so, is even more extensive than the findings of the research itself. The extent of this kind of knowledge is a function of the numbers and kinds of interactions that enable mutual discoveries to occur among customers and vendors.

A PROFILE OF EXTRAMURAL VENDORS

DoD's Directorate for Information Operations and Reports (DIOR) annually compiles a set of annotated tables denoting the expenditures for R&D during each previous year. From these we can derive a profile of the vendors. Abstracted from databases that are derived from contracts, the tables are grouped according to the defense components for which the money was spent. The data include dollar sums for prime contracts of more than $25 thousand, and they are displayed in various ways: by type of procurement program, by competitive awards, by contract pricing provisions, and by types of contractors. The tables provide baseline information for the preparation of ensuing budgets. Used for monitoring, they illustrate orders of magnitude, and they enable comparisons. They can also be unpacked to reveal a profile of the R&D vendor community.

The tables for FY 1991 offer a reasonable historical view of the vendor community. Prior to that year the roster of vendors had been fairly constant for a decade or more, but thereafter, the defense drawdown began to cause major realignments. A front-page story in *The Wall Street Journal* (Ricks, 1993), for example, recounts changes in the market that were occurring as a result of the defense drawdown. Mergers and buyouts among the vendors began to change the databases, but in 1991 these changes had not yet occurred. So the FY 1991 expenditures for R&D, which were shaped in the year preceding, offer a reliable picture of the contracting community that had evolved during the Cold War. Three tables pertaining to R&D vendors display different scales of information for 1991. Compiled by DIOR, the *Department of Defense prime contract awards* (DIOR, 1991c) is updated semi-annually and covers a five-year period. The other two, updated annually, are more detailed: *100 companies receiving the largest dollar volume of prime contract awards* (DIOR, 1991a) and *500 contractors receiving the largest dollar volume of prime contract awards for RDT&E* (DIOR, 1991b).

The different data tables yield a composite picture of DoD's external contracting for RDT&E during a 5-year period. From 1987 through 1991, DoD agencies and activities awarded 2.7 million prime contracts, totaling

$21 billion. The proportionate expenditures for goods and services are displayed in Table 7.1 according to the budget categories.

Although categorically explicit, this representation does not account for the domain of generic research. The spending for DoD's technology base (categories 6.1 and 6.2) is exactly recorded. But the all-important distinction between Advanced Technology Development (6.3A) and Advanced Systems Development (6.3B) is not displayed. This gross reporting appears to correlate the actual use of the funds. As noted earlier, the boundary between 6.3A and 6.3B is indistinct; money passes back and forth between them.

The tables yield much more information for the vendor community. The prime contract recipients in FY 1991 represented three orders of magnitude: several thousand business corporations, several hundred nonprofit organizations, and several dozen federal agencies and foreign vendors. Among all U.S. vendors, 87% of the dollars went to business corporations; 12% went about equally to educational institutions and other nonprofit organizations; and 1% went to other federal agencies and foreign vendors. Among the 500 leading RDT&E contractors in all categories, prime contract awards to any one vendor ranged from $1.7 billion down to $1.7 million.

The dollar distribution was markedly disproportionate. Of the $21 billion awarded to the top 500 contractors, the top four (Martin Marietta, General Electric, Grumman Aerospace, and McDonnell Douglas) received 25%. The next 10 commercial vendors received another 26%. The remaining 49% went to the other 386 vendors. The fall-off was equally steep among educational institutions. Slightly over $2.5 billion was awarded to 82 colleges, universities, university systems, and other nonprofit organizations. But 31% of that amount went to only two institutions (Johns Hopkins and MIT). The implications of this demographic pattern could be the subject of yet another study of the military-industrial complex.

This set of figures represents only the point of the spear. Those 500 top winners were only a modest fraction of some 4,000 R&D vendors that

TABLE 7.1

Budget Category	*Proportion of Budget Awarded*
6.1 Research	5.1%
6.2 Exploratory Development	10.9%
6.3 Advanced Development	30.1%
6.4 Engineering Development	27.8%
6.5 Management and Support	5.4%
6.6 Operational Systems	20.6%
Total Contract Awards for RDT&E	99.9%

received prime contracts. Not included in DoD's census tables are the myriad subcontractors and sub-subcontractors located in virtually every congressional district in the nation. The extramural market for defense R&D is immense.

By listing the separate contracts awarded to each vendor, the data tables give some indication of the various vendors' organizational structure. That structure tends to be generic. Although commonly referred to as laboratories, R&D organizations are more accurately described as laboratory clusters, wherein each laboratory has its own assigned functions. In both profit and nonprofit organizations, each individual laboratory is a separate department. Although some activities might overlap, each laboratory has a relatively specialized mission area, which is likely to be interdisciplinary. Each laboratory has a staff of specialists whose work pertains to that area. They are typically engineers and scientists who represent related disciplines or subdisciplines and who are accustomed to being team members. Each director is expected to justify the laboratory's use of resources in its mission area. This responsibility entails developing knowledge products related to some recognizable goal.

In general, each laboratory's goal is knowledge output in a form that will assure continued funding; it must produce a *quo* for its *quid*, so to speak. The output will lie among a range of possibilities that characterize basic and applied research. It may include analyzing some set of phenomena, formulating a set of related empirical problems, or whatever else might satisfy a perceived need for knowledge. Depending on the overall mission of the vendor, each laboratory's output is within the boundaries of the three defense R&D budget categories: 6.1, 6.2, and 6.3A. Above all, the laboratory's management must be capable of selling the output to some agency or activity within DoD.

For most of the academic laboratories that deal with DoD, the exchange of knowledge for money is conducted on the basis of reimbursed costs. It primarily entails research leading to findings that will pass peer review and qualify for publication in learned journals. The exchange also includes some professional training and subsidy for graduate students. Such research is primarily in the category of 6.1, and the funding is primarily in the form of grants. The grant of money is based on a simple, open-ended agreement that research will be conducted in a given field of endeavor over a given period of time, with the understanding that the primary beneficiaries will be the principal investigator and, theoretically, the public. The official rationale for 6.1 funding includes both the general public good and the possibility that research output might serve some future military objective.

A special category of R&D vendors, often lumped with educational institutions, consists of nonprofit organizations that perform continuing services for DoD. These are mainly Federally Funded Research and Development Centers (FFRDCs). Their services are dedicated to sector research in designated fields of science or engineering, or operational support in public policy, logistics, or management. The FFRDCs have a unique relationship with their defense customers, as legislated by Congress. Although they exist to serve those customers, the intent is to distance the FFRDCs from pressures that might jeopardize the independence of their findings. Depending on the nature of the research in individual cases, the vehicle for sponsorship might be either a grant or a contract. A background paper on the FFRDCs (OTA, 1995) analyzes their changing relationship with the military services since the end of World War II.

But most of the laboratories that deal with DoD are commercial. In their cases, the *quid* for the *quo* entails the sale and purchase of research dedicated to specific and limited objectives. The research is funded through budget categories that have historically been 6.2 and 6.3A, and the primary beneficiary is understood to be the purchaser. Both the *quid* and the *quo* are explicitly bound by contract, but the nature of the R&D is broadening.

A relatively recent, and important, class of defense R&D vendors operating for profit consists of technical service firms. Known generally by the acronyms of their corporate names—such as SAIC, BDM, and PRD—they are relatively unknown to the general public. They deal in information, with the purpose of monitoring or expediting huge weapons contracts and other large commitments of DoD. These service firms employ economists, engineers, computer scientists, physicists, chemists, and geologists. Their specialists might include soils engineers, oceanograhpers, environmental engineers, or hazardous waste experts. John Mintz (1993) summarizes their work. They typically "specialize in massaging intelligence data—such as written text, photo images and electronic signals—and repackaging them for new uses" (p. 32). Gathering and interpreting information, they are intensely engaged in knowledge production.

THE MARKET CONDITIONS

Most of the research that DoD funds is purchased research, and it is specified by contract. In these circumstances, the conditions that determine the exchanges of money for research are essentially constraints. Some of them are typical of any market, such as the units by which the product is packaged and sold and the kinds of agreements that bind the sale. Other conditions in this particular market are peculiar to public procurement; they are imposed by the government, and they concern the regulatory codes that govern competition for public funds. Still other limiting conditions pertain to any R&D market; they concern proprietary information and the ownership of intellectual property. All of these conditions exist in this particular market, but the ones that most affect it are the convention of dividing R&D into manageable units for sale and purchase, and the regulatory constraints that shape the interaction of customer and vendor.

THE PROJECT

The basic unit for sale and purchase in this R&D market is the project. Sponsored by the customer and performed by the vendor, the project is a set of related tasks, undertaken to meet certain objectives within a specified budget and a designated period of time. The projects are commonly related to the defense critical technologies, and the tasks in each project are explicitly defined by a Statement of Work (SOW). Tasks are commonly interdisciplinary; they involve subdisciplines of physics, engineering, computer science, and electronics in various combinations with mathematics and chemistry. And the project, with its included tasks, usually fits within a larger R&D program. By its nature, the project is instrumental to some broader set of objectives.

Such local pragmatism is typical of all government-sponsored R&D, Nathan Reingold (1991) observes; it is a kind of limited instrumentalism. He sees in it an opportunistic use of theory, data, instruments, and techniques to solve problems, but with little or no concern for grander conceptual structures. Reingold is accurate about the defense R&D project. Its limited nature conveniently accommodates both customer and vendor. Given the customer's limited budget, its mission priorities can best be factored into discrete and successive projects. This mode of purchasing R&D is essentially what determines the vendor's organizational structure—the laboratory cluster. Staffed by specialists in different fields of inquiry, individual laboratories provide a flexible capability. Specialists can be assembled in temporary teams to meet the objectives of each new research project.

THE CONTRACT

The project at hand, whatever it happens to be, is the focus of R&D activity. The most comprehensive description of it is, surprisingly, not the final report, but the contract. Because it binds the customer and the vendor to terms, the contract is the document that completely delineates the project. It specifies the project's scope and objectives; the work to be done, with measurements of its quantity, quality, and deliverable outcomes; the price and payment terms; and the duration of these related activities. In the intensely competitive environment of the procurement process, most project initiatives die along the way or get subsumed by some other effort; they do not ever arrive at a contract. But whether or not a proposed project is funded, all documents pertaining to its formulation, issued by customer or vendor, are instrumental to a presumed contract.

In any case, whether or not a contract materializes, the negotiations that anticipate a funded project are expensive for both parties. Even after all the agreements have been reached, it may take many months before the work is initiated. During this extended period, there is always a possibility of the project being scratched; Congress might withdraw the money, or some new program might take priority. So the defense agencies and activities always look for alternative ways to minimize the contracting process. If they could avoid altogether the burden of maintaining free and open competition among vendors, as mandated by law, so much the better. Like any other customer, they would prefer to buy from a supplier whom they know and trust.

One way of eliminating competition, called "sole-sourcing," has bearing on the flows of information between customer and vendor. This alternative depends on invoking exceptions to the laws protecting competition, so as to permit the selection of one source only, one vendor, to perform the intended research. It was commonly practiced for some 30 years until Congress restricted its use in 1984 in the Competition in Contracting Act (CICA).[1] During that long interval, customers and their preferred vendors could engage in the free exchange of information, incubating ideas and nurturing them into projects. Reciprocal iterations of a problem would naturally generate new inferences, new knowledge.

After the restrictions levied by CICA, military customers began to invoke other contract vehicles, one of them being the indefinite quantity contract (IQC). Normally used with equipment suppliers, it has been modified for R&D purposes. The IQC is an umbrella contract that provides for lesser contracts to be added later. The modified version, known as a task-ordering contract, preserves certain advantages of sole-sourcing, as regards the flow

[1]CICA limits sole-sourcing to any one of seven enabling conditions: there is (a) only one responsible source, or (b) an unusual and compelling urgency; or there is a need (c) to maintain an R&D capability at an educational institution or FFRDC or (d) to comply with a statue or (e) an international agrement or (f) a cabinet-level confirmation of a special need, or (g) to avoid a situation where full and-open competition would compromise national security.

of information between customer and vendor. Deliberately open-ended, task ordering provides for the subsequent formulation of discrete but related projects. The prior contract qualifies a given set of vendors to perform possible tasks that a customer may want to purchase; it is followed later by specific task orders to those designated vendors. Separately negotiated, each subsequent task order becomes a project. The process of procuring this kind of arrangement involves ongoing communication among the customer and several vendors, and it bears on the incubation of new projects.

Task ordering in research projects is favored by customers and vendors alike for its cost saving and its predictive quality. But there are caveats, as Peter Ritenburg (1988) explains. The process has been questioned on the basis of its limiting full and open competition. It has also caused evident concern in the GAO over how it should be regulated.

No matter what the contract vehicle turns out to be, however, there are two principal players in the development of a project. One, representing the customer, is the contracting officer's technical representative (COTR), often called the technical monitor. The other, representing the vendor, is a principal scientist or engineer who has been designated as project director. In the negotiations that progressively define a project, the chief task of these two persons is to accommodate the different objectives and priorities of their respective organizations in arriving at a set of feasible tasks that are mutually compatible. In their sharing of information and in their mutual development of substantive ideas, cooperation is their principal mode of interaction. Project design entails group efforts on both sides of the table; bringing those separate efforts together and negotiating the project is the joint responsibility of these two principals.

But one cautionary detour is appropriate. Negotiation is a broad concept; it connotes polar meanings, from cooperation at one extreme to adversarial bargaining at the other. Which meaning is operative depends on who is doing the negotiating and in what situation. The technical monitor and the project director together can delineate the substantive aspects of a possible project, but neither of these persons can authorize a contract. That latter responsibility lies with a different pair of players: the contract officers who respectively represent the customer and the vendor. The orientation of these two persons is toward the contract rather than the project. The customer's contracting officer has very different objectives from those of the COTR who has evolved the project with the vendor's project director.

The role of the customer's contracting officer is about the same as the role of an insurance claims adjuster—to save the company money and keep it out of litigation. It is a prominent role in government contracting, and it provokes vendors into passionate generalizations about having to do business with a Jekyll-and-Hyde customer; it courts them on the one hand and distrusts them on the other, both in the name of public welfare. The customer's contracting officer is only one of a multitude of biting insects that the vendor encounters in the swamp. Defensive audit procedures and adversarial relationships levy costly compliance rules, all in the name of free competition.

To vendors, the customer's contracting officer is the first symptom of all this adversity, interfering with a project by negotiating over a whole new set of objectives. Those objectives include saving money, insisting on logistical details, and policing possible conflicts of interest. They might even extend to subcontracting arrangements or to the political implications of a laboratory team's constituency. In turn, the vendor's bargaining representative is often burdened with educating the contract monitor about R&D, explaining the differences between a knowledge product and a carload of socks. Gratuitous constraints do not serve the vendor. So they negotiate. And their staffs confer with them. Vendors are fond of observing that the number of auditors associated with a project often exceeds the number of people who will perform the R&D.

The point of this detour about negotiation is to make clear what knowledge production is *not*. Contractual bargaining may be urgent, but it has nothing to do with the flow of substantive ideas in the development of a project.

THE REGULATORY ENVIRONMENT

The R&D contract is the legal record of the goods to be exchanged: the funding and the deliverables. In one sense it is a unique document. Retrospective and cumulative by nature, it incorporates into a single text the results of many months of negotiation, essentially between two scientists or engineers who represent the customer and the vendor. But in another sense the contract is also a generic document; it institutionalizes that agreement by clothing it in a context of regulatory constraints. Every contract is freighted with boilerplate restrictions. They include contract clauses excerpted from the encyclopedic Federal Acquisition Regulation (FAR), that governs all federal contracts. They include codicils to the FAR, which the funding agency has formulated. And they include mandatory certifications of the vendor's intent and status regarding possible conflicts of interest. These clauses and certifications reflect layers of social legislation pertaining to small business, women and minorities, government property, safety regulations, and environmental protection.

The burden of these restrictions is the burden of exquisite accountability. Its symptoms are exhaustive record keeping and reporting procedures, and also the swollen administrative bureaucracies that are needed to cope with them. The burden is cumulative and strangely democratic; it falls on all parties. It is also unique to this market. Explaining the need to minimize the burden, John Deutsch (1993) has drawn a simple analogy to the commercial market. "When we buy a car from GM," he says, "we expect it to work. We don't expect GM to provide us [with] details on how they costed everything. If we don't want the GM car, we go somewhere else" (p. 9).

At about the time Deutsch was making plans to reduce the inertial impact of these restricting regulations, the Army Materiel Command (AMC) authorized a study of its own cost-accounting procedures. The AMC's study was undertaken by the accounting firm of Coopers & Lybrand, and based on a survey of contractors. Summarizing the results, Alfred King (1993) cites the major complaints from the field:

- The aggressive posture towards contractors by the Defense Contract Auditing Agency (DCAA), which is in turn "subject to scrutiny by the General Accounting Office (GAO), the Inspectors General, and sometimes Congress" (p. 48).
- The unique costs of complying with the procurement regulations, which are required by statute; and
- Inconsistent cost accounting, with one set of overhead rates and depreciation for intramural laboratories and another set for extramural contractors, causing mutual distrust when these vendors compete with each other.

King analyzes the substantial costs of the auditing requirements incurred by the contracting agencies as well as the contractors; he calls attention to the intense interaction as "a perfect example of nonvalue activity," and he points to a sobering cause. "The underlying assumption," he says, "is that contractors cannot be trusted" (p. 50).

Complaints abound from all sectors of the vendor community about the official loss of trust in their efforts. Speaking from the university's perspective in a *Science* editorial, Philip Abelson (1991) summarizes the dismay over "the long-term course of evolving deterioration" in the interface between government and research institutions:

> In the early days after World War II, there was a high degree of mutual trust and an absence of bureaucratic requirement. Scientists had freedom to formulate and conduct their programs of research. Later the bureaucrats took over and placed emphasis on project research with highly detailed budgets and detailed research proposals. That, of course, is the road to pedestrian research. (p. 605)

From an industry perspective, Ian Ross (1993), former president of Bell Laboratories, speaks of the unfortunate tendency in the federal government to keep industry at arm's length. His careful observation singles out not so much the restrictive law as its restrictive interpretation:

> Conflict-of-interest laws are usually cited as the reason for this separation, but I suspect we have gone too far in their application. Perhaps we have turned prudent safeguards into an ideology that institutionalizes division. (p. F13)

As these testimonies indicate, federal regulations distance the vendor from the customer. A major factor in contracting for defense R&D, these regulations have significantly shaped the marketplace by creating a rhetorical environment of divisiveness, adversarial bargaining, and lack of trust. Yet the market force prevails; the customer is committed to a purchase, and the vendor will receive the money up front. So the powerful need to exchange counters the antipathy that the regulations cause. At the operational level, the customers' technical representatives and the vendors' project directors manage to work around the restrictions. Their efforts are symptomatic of the dynamics of the R&D market. To abide by the regulations and still manage to do business with each other, the technical specialists who represent the customer and the vendor have evolved a rhetorical coping mechanism. As the next chapter explains, that mechanism is marketing.

INTERACTIVITY AMONG CUSTOMERS AND VENDORS

One other factor that delineates the marketplace for defense R&D is a set of conditions wherein customers and vendors recurrently change their relationships to one another. Such flexible arrangements are limited to individual projects; they can occur within the customer community and within the vendor community, as well as between customers and vendors. And they profoundly affect the exchanges of information and knowledge.

Shifting relationships within the customer community have already been noted. Some of the R&D money allocated to DoD's intramural R&D activities is exchanged among them for services rendered. The lateral flow of money among the R&D activities is also true of their extramural contracting; as a result, the actual end user of a knowledge product is not always evident. As customers, several in-house activities might pool their resources, or one activity might borrow money from another, to be repaid in the next fiscal year. Depending on their different teaming arrangements, one activity might fund a contract for a given project, but another might supervise the competition for the contractor. This latter option, for example, describes ARPA's arrangement with the military agencies.

There are yet other kinds of arrangements within DoD's procurement system that link the parties differently. To explain them requires a temporary change of syntax. So far, it has been convenient to regard the customer and the vendor as generic entities, and to refer to them in the singular. But to illustrate a point, suppose the population of the marketplace were now expanded to include two actual customers and two actual vendors. Even that small population would reveal a geometric increase of possible relationships among them.

To illustrate the permutations that can affect the flows of money and knowledge, imagine the possible combinations in a population of two customers and two vendors. There would be three dyads:

1. customer-with-vendor,
2. customer-with-customer,
3. vendor-with-vendor.

For any one R&D project, each of the three dyads provides two possible relationships: the paired parties might be affiliated, or they might be in competition with one another. So there are six possible juxtapositions.

You could build a figure to illustrate the complexity of these juxtapositions by grounding them to six typical situations in the marketplace. You might annotate the figure with a legend that numbers each paired party (#1 and #2). Then you could supply a pair of symbols, representing either a collaborative or a competitive relationship between them. The result would be a model of dyads and their polar relationships. Then, representing actual situations, such a figure would display six different relationships that are reasonably common in this marketplace (see Table 7.2).

TABLE 7.2
Dyads and Their Polar Relationships

Three Dyads	*Symbols*
Customer #1 and vendor #1	C#1 and V#1
Customer #1 and customer #2	C#1 and C#2
Vendor #1 and vendor #2	V#1 and V#2
Two Operational Relationships	
Affiliated or collaborating	←→
Segregated or competing	→←

Customer and Vendor

(1) A customer retains a vendor, as a consultant, to help prepare technical specifications for a prospective project (C#1←→V#1).

(2) A second vendor responds to the funding announcement of this project by submitting a proposal (C#1→←V#2).

Customer and Customer

(3) One customer initiates a project but assigns to another customer the process of selecting a vendor (C#1←→C#2).

(4) Two customers have similar missions; they initiate projects with similar specifications. Located in different military services or in different congressional districts, they compete for funds (C#1→←C#2).

Vendor and Vendor

(5) Two vendors team up to offer a joint proposal for a project (V#1←→V#2).

(6) These same two vendors compete with each other in separate proposals for a different project (V#1→←V#2).

As the pairs of alignments indicate, any research project might entail very different organizational relationships. But this figure is only an abridgement of the many possibilities, since the permutations are not limited to pairs. A given R&D project might readily engage three parties from two dyads, thereby involving one of those parties in two polar relationships. For example, one customer might compete with another customer over a project while affiliating with a vendor in the same project. Or, conversely, one customer might affiliate with another customer in a project while bargaining with a vendor about it. If we relate these possible combinations to the actual order of magnitude, the true complexity of this marketplace becomes clear.

My point is not to quantify all the possible operational relationships in the marketplace, but rather to suggest the kinds of patterns in which information flows through it. Customers and vendors are continuously repositioning themselves for each new project that engages them. DoD's census of R&D vendors, summarized earlier, is based on 2.7 million R&D contracts in a single year. We can assume that at least that many R&D projects are ongoing every day. Given this order of magnitude, the odds on any one organization, whether customer or vendor, assuming multiple postures at any time are overwhelming. Repositioning one's organization in the marketplace, is continuous and commonplace.

The shifting of institutional roles, of course, conditions the behavior of the technical specialists who represent those institutions. They develop appropriate communication skills. The vendors' project directors and the customers' technical monitors readily understand the perspectives of their counterparts. Although mutually bound by parietal rules intended to keep them at arms length, they know the steps in the dance. Not only are they used to changing partners, they also know how to lead or follow the partner they happen to be with.

Such knowledge sharing is characteristic of an R&D community. The participants are mission oriented, and they are accustomed to working in teams, both within and beyond their own organizations. They are one another's consultants; they share large stocks of knowledge, which they readily augment by exchange. A frequent maxium in the community is that R&D is a people business. The archetype of this operational mode was probably the Manhattan Project. And what has evolved from it is a highly refined version of the collaborative mode.

8

Marketing and the Coproduction of Knowledge

Sought knowledge is a function of two phenomena: disseminating information and interpreting it. Both of these phenomena are prevalent in the defense R&D community. In the several previous chapters we have seen the broad patterns that govern the disseminating of information and the opportunities these patterns provide for interpreting it. They are all permutations of the customer–vendor relationship. In the defense R&D community, that relationship is formulated to a high degree by marketing. In the transactions for procuring R&D services, marketing channels the dissemination of information and generates opportunities for interpreting it. This chapter examines the ways in which marketing operates in the procurement of R&D.

Downstream from where the more obvious political tributaries enter the market, the R&D acquisition process enters its procurement phase. By the time this phase begins for the typical R&D project, the earlier stages of acquisition planning have been completed. Technological needs for a system or subsystem have been confirmed and partially defined. Congress has appropriated the funds for R&D, and the money has been transferred to a DoD activity for the procurement of designated tasks. Formal competition among vendors is about to begin.

But long before this point in the planning, another process has been going on. Some contracting officer's technical representative (COTR) at that DoD activity has been gradually defining relevant technical problems in dialogues with a number of possible vendors. And those dialogues have incidentally revealed the capabilities of these vendors. This is the context within which marketing occurs. Customers and vendors need each other;

their mutual dependency is continuous. The vendor needs the customers in order to stay in business. And customers have comparable needs; to justify their own continued existence, they need their suppliers to succeed in their assigned tasks. Fruitful R&D projects contribute to the mutual success of both parties.

Customers and vendors have another reason to cultivate each other. They work in a restrictive, regulatory environment that tends to separate them from one another. Adversarial posturing and divisiveness is not uncommon in the procurement process. To get done what needs to be done in designing an R&D project, there is need to counter a pervasive institutional insensitivity. So the technical specialists who represent both customers and vendors have taken advantage of marketing as a coping mechanism. Their operational mode rests on deductive logic. Both parties have compatible objectives, both are constrained by regulations. So they engage in marketing.

In the many analyses of the military-industrial complex, not much has been made of marketing. Yet some comprehension of its immense currency since World War II is critically important to an understanding of knowledge production. Understanding how marketing catalyzes the flows of knowledge in the defense R&D market depends on understanding what marketing is and how it works. So this chapter begins with a summary of marketing and its evolution in the United States economy during the past 50 years. The summary anticipates an explanation of how marketing has modified defense R&D procurement procedures and, thereby, how it has affected knowledge production.

MARKETING AND ITS EVOLUTION

Marketing is generally understood to be a vendor's strategy for creating an environment that will enhance sales. It moves a product into the customer's frame of reference by presenting that product in a recognizable and convincing context. On the back of the flour sack of Pillsbury's Best is a prize-winning recipe for baking. Although the recipe is a sales device, it did not exist until Pillsbury had first organized the baking contest that yielded an array of winning recipes. That contest was a marketing enterprise. On the back of another flour sack, one for Gold Seal Flour, is a coupon that General Mills will redeem for a free recipe book that parents and children can use together. The coupon is a market-research tool for learning more about the company's customer base.

These instances illustrate marketing as a strategy for motivating a prospective customer to make a purchase and to regard that purchase as an investment. But a caution is in order about the idea of motivation. It is an elusive idea, and the familiar economic indicators fall short of explaining it. The indexes of consumer confidence, new housing starts, and wholesale prices all confirm purchasing patterns, but they do not address the psychology of why people go shopping. In *The Wealth of Nations* (1776/1976), Adam Smith took for granted that motivation is inherent in the appetite for trading, and need not be elaborated. He saw it as simply being human: "nobody ever saw a dog make a fair and deliberate exchange of one bone for another with another dog" (I, 2, p. 17).

The idea of motivation is not entirely clear. It is possibly tangential to operant conditioning, wherein a creature's behavior comes first, and then conditioning reinforces it by some means. But the psychology of motivation is not our subject. Insofar as marketing motivates a customer, it does so by reinforcing inherent needs or desires. This seems to be its point of origin.

Marketing is a vendor's strategy for increasing the volume of sales by nurturing the customers' needs or desires. Admirals want ships, generals want tanks and airplanes, and members of Congress want jobs for their constituents. Before the nurturing of such desires can occur, however, the vendor must find out precisely who needs these things and what exactly they want. This is the first step in marketing; it entails a discovery process called market research. The next step is to devise some means of accommodating those needs. Finally, the vending service must be organized or reorganized so that every one of its functions, from product design to product delivery, is based on the customer's needs and desires. Given all these necessary implementations of a central concept, marketing is more complex than most people think it is. It is not merely advertising, and it is not even merely selling.

All customers are alike in one respect. Institutional needs, like human needs, tend to be generalized or highly abstracted, and sometimes not even articulated. But products are highly specific. So marketing is basically a strategy for linking the customer's abstract need to the marketer's specific product or capability and then clarifying this linkage to the prospective customer.

Educating the customer is implicit in this process; teaching is part of the vendor's role. Nearly 100 years ago, John D. Rockefeller taught the Chinese people how to light their homes with oil by first sending them free lamps. In a more contemporary setting, sales persons from drug companies teach doctors about the effects of the medications they sell. A horticulture catalogue tells readers about the life cycle of a fungus gnat, when to plant cover crops, or how to improve composting procedures. Use of this information is made easier by purchasing the vendor's products. A tool catalogue explains how to make jigs to expand the usefulness of power tools or how to design a dust-collection system with components that the vendor can supply. Such catalogues perform one of the historical tasks of trade publications, which is to educate the customer.

R&D vendors served this educational function for many years during the Cold War. Program managers in DoD activities were primarily military personnel who were rotated in and out of desk jobs on a regular basis. As field officers, they were not necessarily experienced in managing an R&D program, or skilled in the technology that the program was developing. Since career advancement depended on their achievements in the field, moreover, they were not necessarily motivated to take strong positions for the sake of advancing a program. Program officers had to learn on the job, and in matters as broadly scattered as specifications for project design, risk assessment, and legislation, vendors naturally became mentors. During the 1980s, the selection and training of R&D program managers at DoD gradually changed, but education remained part of the vendor's role. Shrewd program managers still want trusted technical specialists outside of their own organizations.

But marketing has only recently institutionalized the process of customer education. The Pillsbury Company did not always put prize-winning recipes on its flour sacks. Nearly 40 years ago, Robert J. Kieth (1960), Pillsbury's chief executive officer, remembered a time when manufacturing was ruled by production. He recalled the management philosophy of his own company, which had almost no sense of sales, let alone marketing:

> We are professional flour millers. Blessed with a supply of the finest North American wheat, plenty of water power, and excellent milling machinery, we produce flour of the highest quality. Our basic function is to mill high-quality flour, and of course (and almost incidentally), we must hire salesmen to sell it, just as we hire accountants to keep our books (p. 36).

The importance of nurturing the consumer, when it finally arrived, was a new realization of an old idea. It was inherent in Adam Smith's (1776/1976) economic model. Summarizing his indictment of mercantilism, its trespasses on free competition, and its unremitting support of the producer at the expense of the consumer, Smith wrote:

> Consumption is the sole end and purpose of all production; and the interest of the producer ought to be attended to, only so far as it may be necessary for promoting that of the consumer (IV, 8, p. 179).

Smith's idea found its way into our present economy, and then into our military establishment, only after a series of slow changes, beginning late in the 19th century. By then, specialized labor and manufacturing technology enabled efficient production of many goods for which there were ample markets. With high consumer demand, manufacturers put their attention to the procedures of mass production. But demand for consumer goods began to level off in the 1920s, and it dropped precipitously during the Great Depression. Because goods had to be sold, not merely produced and delivered, a primitive view of marketing emerged. But it was largely limited to advertising, without much effort to distinguish between the two. Serious attention to marketing was delayed throughout World War II, when the shortages of consumer goods made it possible for suppliers to sell whatever they had without much effort. With the end of the war, competition once again became a factor, and the nascent concept of *marketing* emerged. Marketing was defined as early as 1948 by the American Marketing Association. Its Committee on Definitions (1960) reprinted its earlier designation of marketing as the performance of business activities related to the flow of goods and services from producer to consumer or user. This rather flaccid concept of marketing was soon to be supplanted by more muscular versions.

THE MARKETING CONCEPT

Although marketing is obviously grounded in the market economy, unlike economics it is not broadly based in theory. Its development after World War II occurred through corporate policies and procedures. The Annual Report of General Electric (1952) is customarily cited to mark the beginning of this development. The report explains the concept of bringing in the marketer at the beginning, rather than at the end, of the product cycle. The idea was to integrate marketing into each phase of the business enterprise.

> Thus, marketing, through its studies and research, will establish for the engineer . . . what the customer wants in a given product, what price [the customer] is willing to pay, and where and when it will be wanted (p. 21).

The consequence of this policy in carrying out the entire business enterprise became explicitly clear:

> Marketing will have the authority in product planning, production scheduling, and inventory control, as well as in sales distribution and servicing of the product (p. 21).

What is remarkable is how accurately the report explains the organizational structure that has characterized defense contractors ever since.

Marketing was here defined as a management policy, not as a theory, and the distinction is significant. Marketing is essentially pragmatic. Its substance as an academic discipline, for example, heavily depends on the case method of presentation. As typically taught to students, marketing is situational, and the situations typically feature management decisions. From textbooks for college marketing courses, one gathers that the theoretical basis of marketing can be glossed satisfactorily in a single chapter.

In five different textbooks published over a period of 19 years (1964–1983) by five different publishers, I found a remarkable similarity in the presentation of marketing theory.[1] The concepts and commentary that the first book codified turned up in the other four as well, in the same axiomatic manner and in more or less the same sequence. This homogeneity is partly a function of textbook publishing, as anyone familiar with textbooks in one field or another readily recognizes. Textbook production is expensive, and profitable publishing depends on maintaining a consistent volume of sales over long periods of time. This situation dictates a predictability in content and a formulaic approach to the subject. Having once found a market, the publisher feeds it with more of the same. Even though they reappear in updated editions every few years with modest additions of new material, mainstream textbooks all continue to look alike.

The inbred similarity of textbooks on marketing confirms the phenomenon that Thomas Kuhn (1962/1970) formulated about the priority of paradigms. He points to the typical behavior of a professional community in the mature stage of any paradigm, wherein the community is guided by recurrent and quasi-standard applications of a core of solved problems. "These are the community's paradigms, revealed in its textbooks, lectures,

[1](1) Schuyler F. Otteson, William G. Panschar, & James M. Patterson. (1964). *Marketing the firm's viewpoint.* NY: Macmillan. (2) William M. Pride & O. C. Farrell. (1977, 1980). *Marketing: basic concepts and decisions.* Boston: Houghton Mifflin. (3) Louis E. Boone & David L. Kurtz. (1980). *Contemporary marketing.* Hinsdale, IL: Dryden Press. (4) Philip Kotler. (1980, 1983). *Principles of marketing.* Englewood Cliffs, NJ: Prentice-Hall. (5) Thomas C. Kinnear & Kenneth L. Bernhardt. (1983). *Principles of marketing.* Glenview, IL: Scott, Foresman.

and laboratory exercises. By studying them," Kuhn says, "the members of the corresponding community learn their trade" (p. 43).

What the textbooks encode retrospectively is a construct called *the marketing concept* that is applicable to all business firms—manufacturers, wholesalers, and retailers alike. The construct features four separate kinds of business activities, pertaining to product, distribution, price, and promotion. These are classes of activities that management is presumed to control. They occur in an environment characterized by forces that management does not control, including legal and regulatory constraints, societal behavior, political and economic conditions, and technology in general. Such extrinsic forces represent both constraints and possibilities in the firm's business activities.

In the construct called the marketing concept, all of these variables are interrelated (see Fig. 8.1). Three of the textbooks represent the relationships in some version of concentric circles, each divided by spokes; their graphic representations look like wheels within wheels. The hub is labeled consumer. Surrounding it, four segments of an inner wheel (which are labeled with numbers) represent four kinds of marketing activities that management controls. They are product design, distribution, price, and promotion. Surrounding these activities are segments of an outer wheel (labeled by letters). They represent forces in the market environment that are extrinsic to management. Variously designated by the textbooks, they pertain to politics, climate, technology, the economy, and other formative events.

This alignment of in-house activities and extrinsic forces is intended to represent the integrative nature of marketing. The idea is to coordinate the segments of both wheels in order to bring them to bear on the consumer.

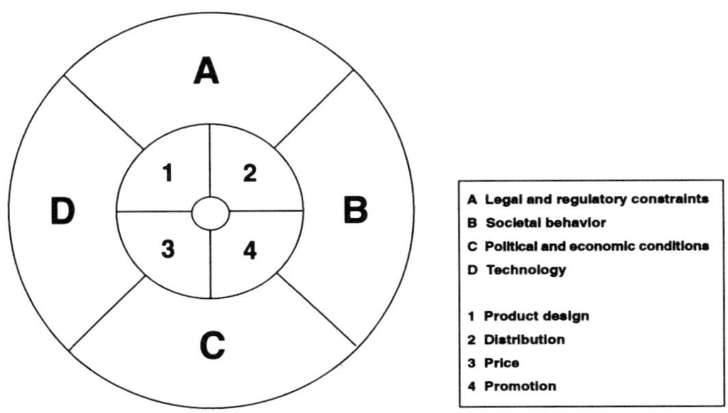

FIG. 8.1. Conventional representation of the marketing concept.

This integrated relationship is what Peter Drucker (1954) had already declared before the first of these textbooks appeared:

> Marketing is so basic that it cannot be considered a separate function. . . . It is the whole business seen from the point of view of its final results, that is, from the customer's point of view. Concern and responsibility for marketing must therefore permeate all areas of the enterprise (p. 38).

THE PARTNERSHIP MODE OF MARKETING

The development of the marketing concept in the procurement of defense R&D has mirrored its evolution in the private sector. That development has forged a new kind of relationship between the R&D vendor and the customer. Based on their mutual attention to the customer's needs, the two parties have learned to form versions of partnership in joint R&D efforts.

The Annual Report of General Electric (1952) is generally regarded as the first milestone in marketing. It reflects the vendor's development of systematic procedures for researching the needs of its customers and then defining its own operations in terms of those needs. Since then, research design for determining customers' needs has been intensely studied, with a resulting—and significant—change in marketing emphasis. What has evolved from corporate engagement in marketing research is the idea of an operational partnership between customer and vendor. Although it was first perceived as a vendor's strategy, market research has increasingly yielded a close working relationship between vendor and customer. As the customer's institutional needs became more complex, both parties pooled their efforts to define these needs and to design systematic ways of accommodating them. Their efforts became joint R&D enterprises.

In this partnering concept, both parties jointly study the customer's entire business organization. Large corporate vendors assemble in-house teams of specialists who are assigned to the customer, who sometimes work at the customer's site. IBM has done this for years in its marketing of mainframe computers, and General Electric has regularly assigned its engineers to customers' factories. *Business Week* (1992) cites comparable arrangements. Flexible teams at DuPont, composed of chemists, marketers, and regulatory specialists, work with customers in new product development. Teams of specialists at Reynolds educate customers on new uses of aluminum. Such commitments to the customer integrate marketing with the total business activity and necessarily change the way in which the vending company is organized.

One reason for accelerating such R&D partnerships lies in the advancement of manufacturing technologies, conspicuously in fields related to computing. Building computer networks with components already purchased but not designed to be compatible is inherently a joint venture between customer and vendor. Advertisements by computer manufacturers and software vendors in business-management publications assert that such endeavors require a level of intimacy that goes beyond conventional notions of cooperation. Such intimate partnership profits both parties. Without adding any employees, customers benefit from R&D efforts focused exclusively on their own goals and constraints. For their part, vendors know that customer retention is their major leverage in reducing the cost of sales. From both perspectives, therefore, interdependence in the R&D effort is desirable. Its advantages have long been obvious in the acquisition of computer-driven weapon systems for defense.

Systematic, institutional marketing evolved concurrently, during the 1950s, with the rapid U.S. military buildup in response to Soviet achievements in space. For the development of defense technologies, Congress appropriated more than 80% of the entire federal R&D expenditure. The Defense Reorganization Act of 1958 created the office of DDR&E to coordinate growing numbers of contracts for R&D. At the top of the management hierarchy, DDR&E provided an interface with extramural vendors.

Interaction was not limited to the upper levels of management. It occurred also at the operational level in defense R&D activities, with respect to the protocols for procuring R&D. Soon after the defense buildup began in the 1950s, directed dialogue between program managers and extramural vendors became commonplace.

MARKETING AND THE PROTOCOLS FOR PROCURING R&D

Evidence of the tight relationships between customers and vendors was confirmed by some remarkable findings of a set of studies conducted by MIT. Shortly after the establishment of DDR&E, a research team at MIT's Sloan School of Management conducted three separate surveys of DoD protocols in the purchasing of R&D. Edward Roberts (1965) describes the surveys as having two purposes: (a) to identify patterns in the R&D procurement process, and (b) to assess the patterns from the perspectives of both the federal government and the contractors. Although the three surveys analyzed different data sets, they all confirmed a distinct pattern of collaboration between customers and vendors. In retrospect, the importance of the surveys is the durability of their findings in the years that followed.

Two of the surveys were based on data gathered from two DoD field centers during different two-year periods between 1960 and 1964. Interviews based on the data were then conducted with the COTRs who had originated nearly 100 different R&D projects during this extended period; these same persons had also supervised the evaluation of the proposals for those projects. A third survey, covering the same period of 1960–1964, focused on the R&D suppliers that had participated in competitive bids for contracts; it highlighted the ways in which vendors' marketing initiatives predisposed the COTRs to the contract awards that they made.

The procurement protocol that was assessed in all three studies occurred in several stages over the period of a year. Within each R&D activity, the protocol began with a procurement request (PR) and an in-house authorization process. Then proposals were solicited from bidders, followed by the evaluation of those proposals by the same engineers and scientists who had initiated the PRs. Among the announced criteria for awarding contracts, technical merit was listed first. This criterion was confirmed in the two studies that focused on the defense R&D activities. Most of the winning contractors were preferred on technical grounds. But the studies also revealed the startling fact that the winning contractors had been preferred many months *in advance* of the bidding.

The findings of these two studies revealed how marketing activities had shaped the procurement protocol. They were based on comparisons of initial PRs and the final contract awards. Each PR had been recorded on a printed form that routinely provided a place for project initiators to list possible contractors for that project. The MIT researchers composed these lists into a database. After purging all alphabetized lists and also lists of suggested contractors that had submitted no proposals, the researchers found remarkably high correlations between the earlier listing order of the possible contractors on the PR and the later contract awards. The contractors that had been designated much earlier as number 1 or number 2 in the listing order on the PRs later won most of the awards.

What could account for such predisposition? To audit these high correlations, the MIT researchers designed a third survey, which focused on R&D contractors. It was a survey of responses to a detailed questionnaire sent to bidders for R&D contracts during the same four-year period covered in the other two studies. The survey was based on an unusually large sample; of the 1,100 contractors that were queried, approximately 40% responded.

This third study focused on the protocol for soliciting bids from vendors after an R&D project had been authorized. The principal instrument for soliciting bids at that time was a Request for Proposal (RFP) that was issued with strict limitations on communication between customer and vendor. Although alternative procedures for soliciting proposals have since been developed, the RFP is still a prominent mechanism in the R&D market, and it is still subject to the same limiting rules of communication. The RFP is announced in the *Commerce Business Daily* (CBD), and the announcement date is a critical point of reference in negotiations. Procurement integrity laws permit communication prior to that date, but not after it, except in public meetings open to all vendors.

Any marketing activities that occur, therefore, must take place *prior* to the RFP's announcement. It is a truism among vendors that by the time you see the announcement in the CBD, it is too late; some other team has already won the ball game. Given this situation, what the MIT surveys revealed was the astonishing degree to which prior marketing activities had evidently predisposed the choice of contract awards. Interaction between the customer's project initiators and the vendor's project directors, before the announcement of the RFP, apparently had significant effect on the evaluation of proposals.

By using contract awards to classify vendors as either winners or losers, the researchers compiled a profile of each category. Not surprisingly, the winners had more prior technical experience than the losers. They had also maintained much closer contact with the technical specialists in R&D activities. Therein lay the distinctive feature of the winners: they were proactive marketers.

The nature of the interaction between winning vendors and their customers was borne out by the vendor's answers to a set of related questions in the attitude survey. The winners were more aware, than the losers were, of the value of personal contacts with COTRs. They knew their names and their responsibilities, and they were in the habit of exchanging ideas with them, both formally and informally. The winners wrote their proposals with these readers in mind, typically designing the technical approach to a problem in ways that would satisfy their readers' known preferences. They even directed the language of their proposals to the individuals whom they knew.

The third survey of procurement protocol explains the findings of the first two. It bears out the efficacy of sustained, close working relationships.

One finding, for example, pertains to the possession of prior information: the winning contractors tended to have *more information* about a proposed R&D project than what the RFP publicly furnished to all bidders. Another finding concerns the cost of proposal preparation: the winning contractors spent *less time and less money*, than the losers did, in preparing the proposal document. Finally, a comparison of winners and losers that bid on the same jobs revealed that the winners had consistently submitted *lower bids*. One inference from these findings—that the winners had more information, achieved a lower cost of effort, and made lower bids—was that the winners' contacts with the project innovators had given them a better understanding of what the customer really wanted, thereby relieving them of the need to cover all problem-solving possibilities in their proposals.

This was the inference that Roberts (1967), the director of the MIT studies, emphasized when he described the results. He pointed out, moreover, that the principal interaction between customers and the winning vendors had occurred *before* the RFP was issued. He underlined the timing of this activity, stressing not only its legality but its efficacy. The interval between the customer's commitment to a project and the RFP's date of issue is when the project's specifications are being designed. And that is when marketing has great leverage. Viewed in these terms, marketing reduces the sales cost. But it does not work randomly; the effort must be continuous.

INSTITUTIONALIZING THE PARTNERSHIP

Since the MIT studies were based on data that are now more than 35 years old, it is tempting to see if those findings still have currency. Actually, any parallel study for follow-up purposes would be unlikely now, because the rules for competition have changed so considerably. The changes have broadened the opportunity for customers and vendors to interact. The Competition in Contracting Act of 1984 was a watershed; it led to new federal procedures for announcing research opportunities. Chief among these was a new protocol based on the Broad Agency Announcement (BAA), also known as the Broad Area Announcement. This instrument actually encourages marketing activities.

The BAA has helped to formalize the idea of partnering. One of the architects of the BAA, Joseph Bolos (1991), explains how it works. With timely publication in the *Commerce Business Daily*, the BAA describes a range of needs for related knowledge products and invites vendors to propose projects within this range. It also specifies a sum of money to be awarded, but without reference to the number of projects that might be funded. And it sets a long lead time, usually a year, for the submission of proposals. Unlike the narrower, more traditional Request for Proposal (RFP), wherein the customer has already decided on a specific project, the BAA invites vendors to construct the problems that they propose to solve. In other words, the BAA invites the vendor to help the customer formulate the customer's own needs more precisely; it establishes a condition for the mutual exploration of ideas. The BAA is a candid invitation to engage in partnering for the purpose of knowledge production.

Legislation has also authorized other specific forms of cooperative research between the private and public sectors. The Federal Technology Transfer Act of 1986 and subsequent amendments have extended to industrial R&D firms the same opportunities that universities have for knowledge exchange with Federal laboratories. Follow-up studies of this legislation document its support among the R&D firms. One survey that focused on the Industrial Research Institute (IRI) pertains more generally to the efficacy of information exchange. This legislation confirms a continuing interest among R&D suppliers in cultivating more interactive relationships with their federal counterparts.

Undertaken in the School of Public Policy at Georgia Tech, the study is broadly based; IRI company members conduct 85% of all industrially-performed research in the United States. The responses of chief technology officers and laboratory heads in about 25% of these companies confirmed their strong approval of interacting with the federal laboratories. David Roessner and Anne Wise (1993) report that when the respondents were presented with multiple options for interaction with federal laboratories, they clearly favored Cooperative Research and Development Agreements (CRADAs). But the newsworthy aspect of the study lies in their reasons for engaging in joint ventures with the federal laboratories.

It had been assumed by the survey sponsors that the new licensing provisions in the legislation, in support of commercial profit, would provide the basis for affirmative response from the R&D suppliers. But their incentives for participation are completely different. The R&D officers in these companies regard the principal payoffs of interaction with federal laboratories to be access to new information, leveraged R&D, and risk sharing. In sum, the attraction lies in the collaborative process itself.

This confirmation of a new marketing environment in federal contracting is compatible with the kind of outreach that DoD has directed its R&D activities to undertake. Collaborative negotiation not only reduces the cost of selling, as the earlier MIT study confirmed; it also reduces the cost of purchasing. R&D activities court suppliers by institutionalizing customer-vendor relationships wherever possible. DoD's agencies and R&D activities spend money to orient suppliers to their needs, quality standards, and business policies. Publications entitled *Selling to the Military* or *Guide to Programs* are commonplace; regularly updated, they are widely circulated. The R&D activities also sponsor training programs. They maintain directories of vendors and regularly update public announcements of the kinds of R&D they need. They publish names and phone numbers of program officers with whom vendors are explicitly invited to discuss their ideas. Such deliberate nurturing helps customers to achieve quality control and cost efficiency.

The R&D activities regularly sponsor scholarly meetings focused on special fields of endeavor in the Defense S&T Program. At these meetings, scientists and engineers representing customers and vendors meet to present papers, discuss program needs, and review current research that might be relevant to these needs. The R&D activities take other initiatives as well. COTRs visit their vendors' facilities, where they closely monitor work in progress. It is not uncommon for a mission activity to award a modest contract in order to assess a vendor's capabilities for some possible future project. These are all means of developing dialogues among interested parties.

THE POSSIBILITIES OF DIALOGUE

The dialogues begin with needs and capabilities: the needs of the customer and the capabilities of the vendor. And they address some problem or other with successive iterations. Well-regarded in the defense R&D community, directed dialogue has become a reliable means of refining technical ideas through progressive problem formulation. Embedded in the market culture and shaped by the notion of partnership between customer and vendor, this essential marketing tool is a major means of coproducing knowledge.

The efficacy of the dialogue was a matter of common agreement among the vendors with whom I talked, although there was one qualification on which they agreed. The quality of the dialogue depends on the people who are participating. If the COTR and the project director have been around for awhile, they can learn from each other and pool what they learn. But there are idiosyncratic situations where real dialogue never happens. Sometimes the COTR is pressed to make closure on a certain project. As a result, the deadline for completion may be unrealistic, or it may be unlikely that the work can be done with the available resources. In such situations, as one project director told me, "You end up with a proposal that you put in the mail hoping you won't win the contract. Yet, there are other times when coproduction happens with amazing results."

Intellectually, directed dialogue proceeds from an intriguing puzzle, which is how to identify what you are looking for when you don't quite know what it is. The puzzle naturally invites some procedure for thinking outside of one's own head. The customer's needs are likely at first to seem ill-defined or distorted or fragmented until two parties with different perspectives can develop a language for defining them and a context for assessing them. Developing a language and a context at the same time entails personal relationships; it also entails aggressive mental procedures, mutually undertaken: searching for patterns, testing analogies, and reformulating problem statements so as to find new common denominators.

When two parties with differing perspectives are engaged in this shared task, each builds on the other's ideas. Their dialogue becomes a dialectic, producing new knowledge that neither party had at the outset. The formal exchange of documents punctuates the dialogue, and the documents themselves confirm the dialectic. As the procurement process moves toward project definition, customer and vendor successively iterate the specifications of the knowledge product that the project is intended to yield. Coproduction of knowledge is the result.

Coming to Terms III

The defense R&D community subsists on words, billions of them daily, and millions of them repetitious. There is a fleeting analogy of this condition in the poem, "East Coker," by T. S. Eliot (1963), where the persona is struggling to get a vast subject said definitively. "You say I am repeating [s]omething I have said before," the voice says. "I shall say it again. Shall I say it again?" (p. 187).

Just how many words are needed to clarify the record on any occasion is a matter of judgment. But in a situation where the speaker's credibility or sense of self is at issue, the number of words is likely to increase; in one's effort to be definitive on so important a subject, the record is likely to expand.

Norman Augustine (1983) caricatures this proposition by observing that the most critical aerospace material is now widely believed to be wood pulp. In the first chapter of *Augustine's Laws*, entitled "Paper Airplanes," Augustine lavishes attention on the volumes of paper required in the preparation of proposals by defense contractors. He devises a new unit of measurement, the *Load Factor*, which is the number of pages contained in typical proposals for any project divided by the dollar value of the project. Then he graphs the Load Factor for 69 actual DoD aerospace projects and interprets the results by various measurements of volume. One bidder's proposals during four rounds of competition for the C-5A cargo aircraft, for example, stacked up to a height of 211 feet. Another contractor's proposals for the Advanced Helicopter Improvement Program exceeded the takeoff weight of the helicopter.

Augustine calls attention to federal customers as well as vendors. He cites one Request for Proposal that extended to 1,200 pages and was later supplemented by a "Clarification Document." And he contrasts this achievement with the output of a more primitive management system. When the Army Signal Corps purchased the development of an aircraft from the Wright Brothers, the Request for Proposal consumed fully one page. The entire contract was two pages long.

We get the point. Augustine has consummately settled the matter of paper volume required in the defense procurement system. But the words that go on the paper merit something more to be said. There is an elaborate body of written discourse that not only records the procurement process, but also enables it to happen. The discourse is formal, technical, and carefully crafted. In delineating the procurement process it reveals patterns of interpreting information, of producing knowledge.

The next three chapters address these patterns. Chapter 8 describes the systemic constraints on this discourse community and the social context in which the principal documents are generated. The documents are transactional in nature. Chapter 9 analyzes the textual structure that prevails in the transactional documents, and the way in which structure is used to interpret information. Chapter 10 describes another kind of document, a marketing tool called the capability statement, which floods the defense S&T community. Chapter 11 describes a fundamental relationship among the textual properties of all these documents. They are knowledge-producing properties.

In observing these properties, the chapters that follow invoke several concepts familiar to discourse analysis. One of the concepts is *argument*, one of them is *context*, and one of them is *genre*. We need to come to terms with these concepts at the outset, the better to use them in analyzing the discourse. So coming to terms with them is our present business.

ARGUMENT AND THE PRODUCTION OF KNOWLEDGE

Arguments are intended to persuade; they are a function of rhetoric. In the accepted historical and current usage, rhetoric is both a process and a structure. It denotes the process of building a persuasive argument, and it also denotes the argument so constructed. In either case, *argument* is the key word. The argument of a text is understood to be reasonable. This is a crucial premise in discussing rhetoric, since street usage is pejorative, associating rhetoric with sophistry. Such usage notwithstanding, any serious discussion of rhetoric assumes that reasonable argument is the basis for achieving persuasion.

One outcome of reasonable argument is the production of knowledge, since, by its nature, any reasonable argument interprets the information that it organizes. In fact, its organization *is* an interpretation, so the process of organizing is knowledge-producing. In our first Coming to Terms, I talked about the composing process as being generative, as being a means of discovery. Composing produces knowledge in the sense that it generates new informational relationships; it is epistemic. Argument formalizes the composing process; it organizes the relationships that occur in composing. It formalizes them in order to support some communicative purpose. To put this another way, an argument produces knowledge insofar as it shapes to some purpose the information that it organizes.

An argument typically shapes information by linking a local instance to a larger domain. Suppose a person is composing an argument (in either speech or writing). Its purpose is to get other people to share this person's schema. To this purpose, argument invokes symbols of various sorts that relate two different domains to each other. The first domain is the person's local situation; the second is some larger domain that rules the perspectives of other people as well. To some degree most written discourse serves this agenda, but argument accentuates it. An argument is a sequence of claims or assertions about a local instance, supported by evidence that helps both the writer and the reader to generalize those claims. Such evidence typically includes instrumental assertions as well; it may include figures, tables, and other graphic representations, or typographical devices designed to create certain emphases. These are all schema-sharing symbols.

Argument necessarily occurs in a social context, and it is largely determined by that context. Anticipating the kinds of argument that the next three chapters reveal, therefore, we quickly encounter this notion of context. We need to come to terms with it; we need to explore the kinds of circumstance that shape discourse in the defense R&D community. One relevant point of reference for this exploration is the context of scientific argument. It is relevant because science and technology are so obviously related in the DoD's descriptions of defense strategy.

THE CONTEXT OF SCIENTIFIC ARGUMENT

There are similarities and differences between scientific argument and the kind of argument that prevails in the defense R&D community. We can use them both to help delineate the kind of context we want to explore. Enough has been written about scientific argument to establish firmly the proposition that it is contextual. In this respect, scientific argument is notably similar to argument in the defense R&D community. In *The Manufacture of Knowledge*, Karen Knorr-Cetina (1981) sees the communicative foundation of science "as a form of discursive interaction directed at and sustained by the arguments of others" (p. 14). Furthermore, such interaction is linked to "a social process of negotiation rather than to a logic of individual decision-making" (p. 152).

Negotiation and discursive interaction are familiar habits in the defense R&D community. But there is a difference in the way scientific communities are built up and delineated. Speaking of science, Marjorie Grene (1985) observes that discourse communities lie within the bounds of established disciplines. She perceives the disciplines to be social enterprises in which individuals are dependent on one another for their ideas and the instruments with which they practice. In *Science as Social Knowledge*, Helen Longino (1990) extends the idea of a community as the site of interaction that produces knowledge. "What is called scientific knowledge, then, is produced by a community (ultimately a community of all scientific practitioners) and transcends the contributions of any individual or even any subcommunity within the larger community." In big science, she contends, the integration of disciplinary subtasks into a coherent understanding of a given phenomenon is "a matter of social negotiations" (p. 67ff).

The contextual nature of argument is similar in the communities of both science and defense technology, but there is an underlying difference in function. The function of scientific argument is to persuade a disciplinary community of the validity of certain empirical findings. Its task is to get the findings accepted as facts. Since the 17th century, the consummate goal of converting empirical results into accepted facts has evolved into a complex rhetorical undertaking. Charles Bazerman (1988) has identified styles of argument dedicated to this goal. What has evolved since the *Philosophical Transactions* of the Royal Society is the concept of a discrete text wholly dedicated to converting empirical results to facts. The generic article in a science journal is such a text. Discussing the evolving stylization of this dedicated text in *Shaping Written Knowledge*, Bazerman analyzes its intricate features. The persona, the selectivity of details, and the nested levels of validation are all intended to present the strongest possible claim. To succeed, the principal claim must be consistent with the style of thought and the expressive habits of the audiences to which it is directed. The rhetorical purpose of this dedicated text is to impart authority to the claim that it makes and, thereby, to forestall opposition to it.

Validating a claim in the scientific community depends on more than a dedicated text to make the claim authoritatively; it depends also on the social process of developing a consensus in support of the claim. The earlier phases of the process consist of peer review, which operates primarily through two procedures: (a) the award of grants to researchers, and (b) the publication of research findings. The effect of both procedures is to regulate the kind of knowledge that results from research and also the cultural authority of that knowledge. In *Writing Biology*, Greg Myers (1990) explores the influences of peer review. In the funding of grant awards, peer review evaluates newly proposed research within the context of prior work in the field. In the publication of results, it forces the author to negotiate the level of the claims that can be made. The higher the level of any claim, the more prior literature it necessarily contradicts, so determining the stature of a claim is a matter of negotiation. To get research results published, an author must accommodate the opinions of the referees and the pressure of the journal editor. A typical outcome is that the author tempers the claim so as to lower its level of consequence. Determining what claims will be allowable in the community, even after publication, continues to be adjudicated. If a claim is controversial, unfriendly reviewers are likely to recast it ironically in a less convincing context of research within the field.

Many of the same textual features mark the scientific research article and the proposal statements of vendors in the defense S&T market. In both cases, argument is rhetorically organized; it accommodates the thought patterns and expressive habits of the intended audience. But the purpose of scientific argument differs notably from the purpose of argument in the defense R&D community. Scientific argument is polemical. Within the system of scientific discourse, contention is not noise; it is part of the signal. In the contention over the status of empirical results, arguments become what Bruno Latour (1987) calls trials of strength. In *Science in Action*, Latour analyzes these narrow political purposes of scientific argument. Advancing a claim to the status of fact entails a continuous process of developing alliances. Latour once told a class I was teaching that science without alliances would be like a cathedral without towers.

Science in Action describes the intricacies of these alliances. There are actual alliances among organizations that have a stake in the claim that is being made. There are also paper alliances, scholarly citations that are intricately arranged in the text of the journal article. These paper alliances are assembled to support portions of the claim that the article's author is making. The larger the alliance of citations, the better. In the politics of science, Latour argues, organizational alliances and paper alliances are judged by the same polar standards; alliances are either stronger or weaker, and networks are either longer or shorter.

The paper alliance is composed by an author who is making claims based on certain findings. It exists in the text of a journal article, in the structure of its argument. The function of the paper alliance is to represent the body of prior literature in a way that supports the author's claims. The argument selects and reconfigures passages of previous texts by other authors to present a relentless continuity of results that make the current claim formidable. In Latour's view, the rhetorical purpose of scientific articles is not so much to win approval in the community as to force opponents to accommodate the claim.

THE CONTEXT OF ARGUMENT IN THE DEFENSE R&D COMMUNITY

On this point, argument in the defense R&D community abruptly parts company from scientific argument. Although heavily contextual in both kinds of community, the purpose of argument differs sharply in the two environs. The defense R&D community is market oriented rather than territorial, and the argument that prevails within it is transactional rather than contentious. The basic reason for the transactional nature of argument in this community lies in the customer–vendor relationship that organizes all the disciplines of engineering. Engineers serve customers, and customers have needs. The relationship between the two parties is based on perceived needs and the resolution of those needs. In the market that results from this relationship, the customer advertises those needs, and R&D vendors respond with claims of ways to resolve them. The literature that prevails in the context of this relationship is a literature of needs and claims. It consists of arguments that formulate needs and arguments that formulate claims of solutions to satisfy those needs.

Two characteristics mark such argument. One is the notable way in which it produces knowledge. Transactional in nature, it entails a reciprocity between customer and vendor. In this context, as I mentioned earlier, customer and vendor can often coconstruct knowledge; the arguments they exchange often form a dialectic. Such interactivity is insistently utilitarian. In the empirical realm of defense R&D, performance is paramount; something either works or it doesn't work, or it may work if some variables are changed. The argument that prevails in the discourse of R&D tends toward this kind of pragmatism; it shapes the successive iterations of a problem that customers and vendors exchange. These iterations successively interpret given information, converting it to knowledge.

The other notable characteristic of argument in this community is its bewildering proliferation. The rhetorical purposes of the parties engaged in such argument are far more differentiated than those in the science community. Although all lines of argument serve the basic customer–vendor relationship, changing circumstances cause organizations to shift roles and change partners within this relationship. Because the dyads are continually shifting, the contentious argument of scientific politics would be counterproductive. Although there is fierce competition in the defense R&D market, it is generally not agonistic; it does not usually affect the quality of argument at the operational level. Since today's adversary may be tomorrow's ally, the we–they mindset serves no reasonable function.

THE GENRES OF NEEDS AND CLAIMS

Based on the dyadic customer–vendor relationship, public discourse in the defense R&D community is generic; it is composed of distinct categories of documents. Within any of these categories, the documents share certain textual features that set them apart from other documents. Each category is a separate genre, and the constitutive conventions and codes of each genre are shared by writers and readers alike. We need to come to terms with the concept of *genre*, because each genre has its function in the marketplace. Its distinctive textual features have extrinsic significance in shaping the way the market functions.

To anticipate briefly the taxonomy in the next chapter, each genre is a class of communicative events, or documents. There are relatively few genres in this discourse community, and each one has a name that reflects its particular function in a well-defined communication system. The documents themselves are all expressions of some corporate persona. As in most corporate documents, the first-person pronoun is absent; there is no "I" in sight. Although the documents in each genre vary somewhat in appearance, their texts share a deep formal resemblance; they are aspects of the same architecture. These documents produce knowledge in two ways. One way, as Machlup (1980) holds, is through their dissemination in the discourse community; this process inherently increases stocks of knowledge. The other way is through the architecture of the texts in these documents; the architecture of each genre is a distinctive means of interpreting information.

Eight of the nine principal genres in this discourse community pertain to the procurement process. They are the formal means of negotiating an R&D

project and of recording what is negotiated. They are transactional genres. Although any R&D project yields knowledge, no matter what its output, the transactions by which that project is procured also produce knowledge. The procurement process is a phased interaction between customer and vendor, wherein the needs of one and the claims of the other are gradually formulated. To be sure, each party has disparate objectives at the outset of the procurement process. The exchange of these transactional documents provides successive iterations of some declared need, thereby moving each party toward a common, shared agenda. Each iteration, each generic document, produces a new increment in each party's stock of knowledge. The next chapter, "The Paper Trail: Transactional Genres," maps this incremental process.

The kind of knowledge that the procurement process yields pertains to project design. Beginning with some declared need, the design of a project typically focuses on formulating a problem. Without a problem to solve, there is no R&D, no project. So the knowledge that flows both ways between customers and vendors concerns the framing of problems. The architecture common to the eight transactional genres embodies the rhetorical mode of problem formulation; the mode engages both customer and vendor. Chapter 10, "Formulating the Fundable Problem," analyzes this architecture and shows how it produces knowledge. It also illustrates how customer and vendor coconstruct such knowledge.

In the literature of needs and claims, the eight transactional genres pertain to some declared need. The ninth genre pertains to claims. The claims are a function of a basic dilemma in procuring R&D. Since each R&D project is a new undertaking, the vendor's ability to accomplish it is unknown at the outset. So the customer must choose a vendor on the basis of the vendor's presumed capability. Not surprisingly, therefore, vendors' capability statements abound in this market.

Actually, capability statements are endemic to most markets; they appear in many forms that we tend to take for granted. The business card, the advertisement in the Yellow Pages, and the corporation's annual report to shareholders all have the same underlying purpose. Their purpose is to position some business entity favorably in the customer's view. They are all the same kind of marketing tool.

The intriguing problem of the capability statement lies in making the vendor's claim credible. The statement is essentially a promissory note; it posits a future condition based on past performance. The odds on its uncritical acceptance are about as low as the odds on the credibility of a weather forecast. Predicting the future on the basis of past performance is, of course, a common dilemma. The stock analyst, predicting movement up or down in some sector of the market, copes with it daily. So does the college admissions officer, predicting a student's performance on the basis of aptitude tests. The oddsmaker's task at a horse race is to discern some past pattern that has enough inertial carryover to justify a certain prediction. The task of the capability statement is of this order; it builds on the past. It must also make an attractive future seem viable.

A vendor's past performance can be readily measured by the quality of previous deliverables, by the professional achievements of the vendor's project directors, their patents, or their journal publications. Past performance can be measured by a vendor's research management. Its capacity to monitor and evaluate its own efforts, its ability to attract appropriate sub-contractors or to team with other vendors all count. The trick is to convey these criterial measurements of past performance in a way that serves the customer's vision of a future. Since the task inevitably entails the interpreting or reinterpreting of information, capability statements continually produce new knowledge. Chapter 11, "The Capability Statement: The Truth but Not the Whole Truth," analyzes the texts of a broad range of capability statements that circulate in the defense S&T market.

For convenience, the ensuing chapters address different aspects of documentation one at a time. They present the transactional genres, problem formulation, and the capability statements as if they were separate files overlaid on a computer screen. But in the R&D marketplace these matters are not all that separated. The genres influence one another; they are all structural parts of a huge metatext. They constitute different, successive aspects of a generic problem formulation that produces knowledge. Chapter 12, "The Knowledge Cycle," presents an overview of this metatext; it summarizes the flows of knowledge that prevail in the defense S&T market, and it relates those flows to a process of knowledge production.

9

The Paper Trail: Transactional Genres

Defense Science and Technology can be described as an activity system. It is composed of interactions among the defense RDT&E activities and also interactions between those activities and the extramural R&D vendors. Those various and shifting interactions typically involve relationships that are rhetorical in nature, ones that entail some manner of persuasion. Within the context of this activity system there is a close connection between persuasion and knowledge production, a connection that becomes more evident as this chapter proceeds.

The common currency of persuasion is argument in one form or another. In the defense R&D community, argument is understood to be reasonable and not spurious, and reasonable argument is generally confirmed by formal documents. In coming to terms with the notion of persuasion and the nature of argument, I talked about how they are related. One's goal in persuasion is to get other people to share one's own schema, one's own representation of something or other. In general, the function of reasonable argument is to relate that personal schema to a larger schema that other persons can share. Making that connection is a rhetorical task, a way of relating the argument to the communicative goal.

Reasonable argument makes claims and presents evidence in support of them. But there is more to it. To be persuasive, an argument must somehow graduate the local evidence to a more authoritative status; it must symbolize the evidence it presents in ways that make the claim generalizable and, consequently, more broadly influential. From the text of a reasonable argument we should be able to see this symbolizing process at work.

From both an institutional perspective and an individual perspective, virtually all of the interactions in the defense R&D community entail relationships that are essentially rhetorical and insistently pragmatic. Honesty of purpose is normally transparent. Whether as authors or readers, community members typically know the rules of reasonable argument. They know that claims must be compatible with facts, insofar as possible. They know what is legally and politically allowable, and within these constraints they know how to present information and how to spin it. Moreover, despite shifting alignments, respective goals are mutually understood, as are the rules of argument. Reasonable argument is the accepted manner of discourse in the R&D community, the accepted way of disseminating information.

One way to map the dissemination of information is to track it through the documents that customers and vendors exchange during the design of an R&D project. Every project concludes a history of the negotiation by which it was formulated; so the documents that precede it mark the flow of ideas between customer and vendor. The documents are instrumental tools; as they are exchanged from one party to the other, they progressively refine the objectives of the project by iterating the tasks that it entails. The iterations become an intertext that progressively shapes the project and defines the kinds of discovery to which it will be committed.

Although they exist in great quantity, such documents form only a partial record of the procurement process. When customers and vendors communicate, the media of choice are conversation (by phone or in person) and electronic transmission. The role of the formal document is to confirm the decisions or agreements that have transpired in those other media. Moreover, many of the written records are private and proprietary and, therefore, largely inaccessible. Yet there are enough public documents to delineate the evolution of most R&D projects and the kinds of knowledge that they produce. Like the glyphs on Mayan architecture, these documents constitute the public record of a culture; they are what we have to work with.

The documents are generic. Used repeatedly from one R&D project to another, they represent key events in the procurement process. In the contract system that DoD employs there are eight generic documents. Each of them serves a distinctive rhetorical purpose in the incremental development of an individual project; it is intended to persuade a specific audience about something or other. In the aggregate, they engage the customer and the vendor in coconstructing knowledge about what the funded project will be and also about the nature of its outcome.

This chapter analyzes the generic documents that confirm the key events in the procurement process. My purpose in following this paper trail is to show how the procurement process produces knowledge *prior* to the undertaking of an R&D project. After briefly noting the genre theory that explains the rhetorical purpose of the documents, we turn to the epistemic, or knowledge-producing, nature of the contract process. Then we examine the documents that enable this activity system to function. Noting their chief textual features, we can assess their effect on the way the procurement system works. What should become evident about knowledge production in the defense R&D community is the intimate correlation of text and context.

GENRE THEORY AND RHETORICAL PURPOSE

As a field of discourse study, rhetoric addresses the ways in which argument functions in persuading an audience. One aspect of rhetoric that has been addressed during the past 25 years is the function of genres, or types of utterance. As they pertain to written discourse, genres are texts that have certain features in common. They share ideological content, they emphasize communicative purpose and social action, and they have a distinctive capacity for establishing and conveying rhetorical goals. Individual genres, moreover, tend to cluster in groups, sometimes called genre systems. All of these characteristics mark the documents that customers and vendors typically exchange during the procurement of an R&D project.

In his *Genre Analysis*, John Swales (1990) has examined the literature of genres and constructed a broad theory of their characteristics and of how they function in a social context. At the core of his theory is the tight relationship that exists among the author's purpose, the genre that expresses it, and the discourse community within which the genre document circulates. Critically important to this linkage is the idea of a discourse community.

Such a community is not necessarily geographic or demographic. Rather, it has a virtual existence; it is a socio-rhetorical community, formed for the achievement of certain common goals. Swales observes, for example, that stamp collectors around the world constitute a discourse community. And his example brings to mind a host of others: the Veterans of Foreign Wars, the congregations of the Baptist church, corporate bond analysts, the neighborhood watch, the meat and dairy lobbies, the class of 1948 preparing for the 50th reunion. These special-interest groups, large and small, political or otherwise, wherein communication occurs, are all discourse communities. This is not to say that discourse communities are homogeneous and harmonious. On the contrary, they are heterogeneous, and they can be contentious. What binds their contentions together, however, is some evident rationale for being a community, some subject in common over which to contend.

Since members of a discourse community know and use the genres that help to achieve their common goals, genres are community property. Swales speaks of the tight bond between the properties of a genre and the shared purposes of its users:

> These purposes are recognized by the expert members of the parent discourse community, and thereby constitute the rationale of the genre. This rationale shapes the schematic structure of the discourse and influences and constrains choice of content and style (p. 58).

This linkage is true of the generic documents exchanged between customer and vendor. The most familiar of such documents in any R&D community are the reports of works in progress or works completed, with pertinent correspondence attached. They are the conventional knowledge products of R&D. They are retrospective in nature, and they tend to be summative.

But reports represent only part of the defense R&D effort. They are preceded by a battery of documents exchanged between customer and vendor during the R&D procurement process. Without these prior documents, R&D would not occur. The documents are formative in nature, and they are generic; their texts literally embody a common rationale, which is to develop knowledge products. In this respect, their shared communicative purpose is insistently explicit. To analyze the generative properties of these genres, we need first to address the process from which those properties are derived. As we are about to see, this process is generative.

THE EPISTEMIC NATURE OF THE PROCUREMENT PROCESS

DoD's system for procuring defense R&D entails many recursive cycles of knowledge production in various stages of implementation at any one time. In the aggregate, of course, myriad R&D projects are always ongoing. Any given project in a laboratory is likely to be one of a cluster of related projects sponsored by different military customers. Within the broad field of radar research, for instance, one project might entail experiments in millimeter wave technology, which could serve some later development of high-altitude surveillance capabilities. A concurrent project at a different stage of development might entail low frequency transmissions for close, underwater surveillance. A third project might consist of testing semiconductor materials for their optical properties. So the laboratory's agenda encompasses a spectrum of projects that are either presently related or might someday be so. A comparable spectrum of projects typically engages any sponsoring activity, which is funding an array of projects in various stages at different laboratories in support of its mission.

The R&D project is the correlative of its own history. It is the result of negotiations that have proceeded in stages by means of a discourse exchange system. Exchanged between customer and vendor, successive documents refine the objectives of a project by iterating the activities that might implement those objectives. These iterations progressively shape the project and define the deliverable knowledge products that the project is intended to generate. On some occasions, even documents that have helped formulate related projects are invoked and drawn into the cumulative record—the intertext—of the project at hand.

The successive iterations in this document exchange produce knowledge that will become part of an R&D project, but they typically yield that knowledge before the project even begins. This evident paradox can be explained. The deliverable output of an R&D project, which the vendor contracts to turn over to the customer, is typically a report of empirical results. The report might describe the result of some phenomenon or the performance of a device or system under certain conditions. The output might be an assessment or a validation design or a set of critical measurements or test results. In any case, the deliverable product is most likely to be the report of a quantifiable and certifiable result; it is what the customer actually purchases.

Much of this deliverable knowledge has already been developed by contract time. It has been developed by raising questions that are both conceptual and operational. Exactly what is to be tested? Why is it to be tested? What procedures will be used? What kind of results should be expected? And how will the tests themselves be validated? These questions must have answers before the customer will invest in the project; the answers will constitute new, instrumental knowledge. Yet the questions themselves also constitute such knowledge. The production of instrumental knowledge prior to an R&D project is inherent in the contract system.

Although the DoD activities spend a billion or more dollars a year for research through grants, most of the extramural research that they buy is funded by contract, as measured both by dollar volume and numbers of projects. Therein lies an important distinction in the DoD procurement system, one that concerns both how and when knowledge is produced. The distinction is the basic differences between the two procedures for funding research. Grants are normally intended for open ended research, whereas contracts specify closely defined projects. In these two situations, the researcher's accountability is entirely different.

Grants are usually awarded to individual investigators, and the awards depend primarily on the peer review of research proposals. In the review process, referees offer caveats and possible modifications in emphasis or in methods for the investigator to heed. The program officer gathers those comments and passes them along to the investigator, and together they arrive at appropriate changes in the project. In effect, the program officer engages in separate dialogues with the referees and with the principal investigator. At the time of the grant award, relatively little of what is to be undertaken has yet been discovered. But a great deal is already known about a proposed R&D project before it comes to contract. The contract process is intended to confirm certain deliverables, certain knowledge products, before the project begins. Contract procedure entails a relatively long period of time to delineate those intended products, the conditions of their delivery, and responsibilities of the customer and the vendor. The protracted process of contract development, of specifying ways and means, proceeds in dialogue between the customer and vendor. Together, they make it happen.

There is a real comparison to be made between grants and contracts regarding the ways in which ideas are incubated and nurtured. Figure 9.1 shows this comparison, highly generalized, in terms of knowledge output.

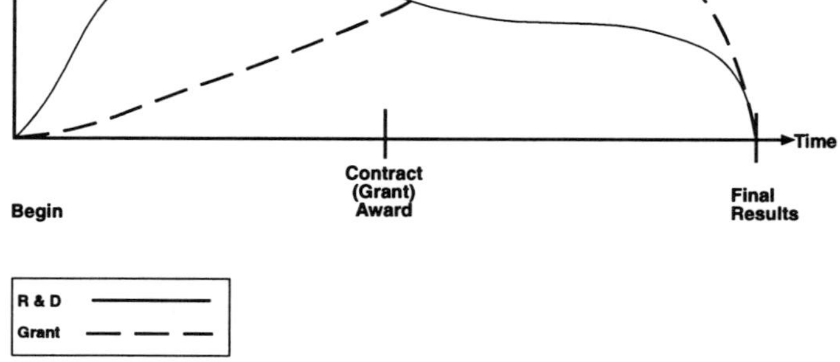

FIG. 9.1. Distribution of knowledge production in the project cycle.

In this model, knowledge producing activity occurs at an earlier stage and at a higher level of effort in contract funding than in grant funding. The model also shows that such activity tails off at a lower level after the contract is signed, since it is devoted largely to confirming or modifying the knowledge specifications that have already been worked out in delineating the proposed project.

Although the comparison is not readily measurable, this model is familiar to research managers. It reflects the distinctive conditions of contract funding. The customer and the vendor are mutually engaged in project definition, and their negotiations occur over long periods of time, lasting many months and sometimes several years. So the principal specifications of the knowledge product that a project should yield have already been made prior to the project's formal beginning.

Interpreting epistemic activity in terms of knowledge products is arguable. Narrowly speaking, a project's intended knowledge product does not exist when the contract is agreed to, because the exact quantification of values remains to be discovered. Yet many components of a knowledge structure by which to formulate those quantitative values have already been established before a contract is signed. Project deliverables, such as assessments, validation designs, critical measurements, or test results, obviously depend on prior specifications. Such specifications include the values to be quantified, the conditions of the tests that will yield those quantitative values, the rationale for such tests and the means of assessing them, and even the parameters of expected results. These components of a knowledge structure will have normally been established before the project formally begins.

THE ENABLING DOCUMENTS

The formidable terrain of the R&D marketplace has already been discussed. It is defined by the Federal Acquisition Regulation (FAR) and its various supplements. These sets of encyclopedic rules require elaborate procedures to insure free and open competition and to provide gateways for new vendors to enter into competition. The insiders know these rules and interact accordingly. The project directors for the vendors, and their counterparts, the technical monitors who represent the various customers, confirm their exchange of ideas through carefully prescribed documents.

A genre map of this terrain consists of pathways that lead from the inception of an idea to the contract for a project. The pathways are different sequences in the exchange of documents between customer and vendor. The documents themselves are coded markers along these pathways. The seven generic documents that move the procurement process along are created prior to any contract. They are transactional in nature, and they have evolved through a long course of adaptation to the market. Four of the documents are initiated by the customer and three by the vendor, and they can be characterized, respectively, as needs and offers. The contract is yet another genre. Here are the eight transactional documents, the first seven of which anticipate a contract:

1–4. **Needs** (Initiated by the Customer)
  Acquisition Forecast
  Sources Sought Announcement (SSA)
  Request for Proposal (RFP)
  Broad Agency (Broad Area) Announcement (BAA)

5–7. **Offers** (Initiated by the Vendor)
  Sources Sought Response (SSR)
  White Paper
  Proposal

8. **Contract** (Agreed on by Customer and Vendor)

These documents are repeated from one project to another, and can be exchanged in different subsets. In the aggregate, they constitute a genre system. Sequentially, in the process of developing a possible contract, the genres are referential and cumulative. The production of knowledge is incremental.

Combining needs and offers, the contract is the most comprehensive genre. It delineates the project and legally binds the agency and the laboratory to its terms. As I noted earlier, the contract specifies the project's scope and objectives, and also the work to be done, with measurements of its quantity and quality. It also specifies the project's deliverable outcomes and the duration of all these related activities. The document itself is a kind of laminated construct. Each contract spells out the terms of a unique agreement, lodged in one of the several layers of generalized information that characterize most other contracts as well.

The contract's structure and most of its content are standardized; a given kind of information is usually found in a predictable place. The text may contain as many as a dozen sections of descriptive material, conventionally located in four parts:

1. **A schedule of conditions** (delineating this particular project and the tasks to be performed).
2. **Regulatory clauses** (excerpted from the FAR and its various codicils and deemed by the customer to be relevant to this project).
3. **Attachments** (copies of printed or published material, included to contextualize the work to be undertaken).
4. **Certifications** (affidavits of intent and corporate status to which the vendor agrees).

Most of this material is boilerplate, extending from possibly 20 pages to 200 or more, depending on the number of regulatory clauses and attachments.

The section of each contract that makes it unique is located in Part A, the schedule of conditions. That section contains the Statement of Work (SOW), which specifies the tasks to be undertaken, the deliverable outcomes that they are intended to yield, and the form in which the outcomes are to be delivered. The SOW is that part of the contract agreement that delineates the project; it is the definitive part of the contract. In addition to whatever else it might specify, the SOW incorporates into the contract the terms offered by the vendor, simply by referring to the vendor's proposal. This incorporation by reference converts two documents, the final proposal and the contract, into one macrotext.

NEEDS AND OFFERS

So much for the contract. Getting to it is the function of the seven transactional documents. Transactional rather than definitive, they provide occasions for presenting or incrementing ideas that might lead to further development. They represent successive iterations of a possible project. Together, the transactional documents constitute a literature of needs and claims, the needs statements initiated by the customer and the claims statements by the vendor. Not all of them are needed on every occasion. Figure 9.2 displays their relationships to one another; it also illustrates the possible pathways, or possible combinations of documents, leading to a final contract.

Two of the documents are substantive. One is the BAA, or broad agency announcement (referred to in the Air Force as a proposed research and development activity, or PRDA). As noted earlier, this document describes an extensive field of possible research that would serve the activity's mission,

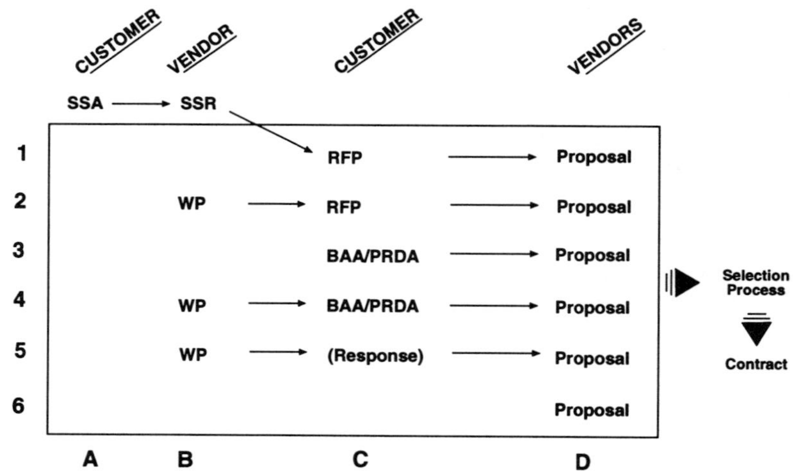

FIG. 9.2. Pathways to an R&D project contract.

and it invites vendors to propose limited projects within that field of endeavor. The other customer initiative, the request for proposals (RFP), also noted earlier, is limited to a project already defined. The RFP includes a SOW that has already been specified.

The RFP and the BAA differently serve the same purpose, which is to invite vendors to help resolve a need (or to engage in an "opportunity"); the customer might choose to use one document or the other. The two documents initiate different responsive procedures, that is, different pathways to a possible contract, and they are likely to occur at different times in an agency's funding cycle. The Broad Area Announcement (or Broad Agency Announcement) occurs early in a longer funding cycle, perhaps 2 or 3 years long, and it signals the possibility of different kinds of projects as yet undefined. In effect, the BAA says, "Here is our general need. Send us your project proposals." The other document, the RFP, occurs in the latter part of a funding cycle, after the customer has conceived of a specific project. In effect, the RFP says, "Here is a project we want to fund. Tell us how you would go about doing it, and how much you would charge."

Because of public concern for free and open competition, the RFP and the BAA are subject to different legislative constraints. They have different rules of engagement, and they offer different kinds of opportunity for dialogue. Historically, the RFP is the older document, and it is much more narrowly regulated to prevent vendors from profiting by inside information. Before an RFP is announced for a given project, any vendor may freely query the customer, but after the date of its announcement there can be no further individual communications until a contract is awarded. Although any vendor may make a query, all vendors must have access to the answer. The formal way around this constraint is for the customer to convene a conference for all bidders to attend, so that they all have the same information. An informal, and more recent, way around the constraint is for the customer to circulate a draft RFP, a preliminary document subject to change, so that vendors' queries and suggestions might help shape the final document.

The strict RFP procedure inevitably creates boundary conditions wherein the system functions erratically. There is a rich folklore about these boundary conditions. Old-timers tell old tales. (They once knew a vendor who advised a customer about an impending RFP, even writing the RFP for the customer, only to lose out in later contract bidding). The tales are legion, but not attributable, of course. Other boundary conditions are less baroque, and they are attributable. Fair and open competition among vendors can be subverted, for example, by the way an RFP is written. Squires (1986) describes how: "Some RFPs are written with a particular vendor in mind. A contract administrator can easily add language to an RFP specifying features that a vendor must display—features almost guaranteeing that only one particular responder can be successful" (p. 130). The RFP might specify that the vendor be situated in a certain geographical area or operate certain kinds of facilities, which place that vendor beyond competition.

After the CICA legislation of 1984 amended many of the irregularities in the RFP protocol, the BAA evolved as a new kind of needs statement, one that presents the opportunity for more open dialogue. As its name indicates, the Broad Area Announcement describes a set of needs in broad terms. It deliberately presents generalized possibilities, thereby inviting queries from vendors and providing for exploratory discussion. The BAA reflects the intention to create occasions for vendors to come forth with their own ideas.

The vendor's formal initiatives are the white paper and the proposal. These two instruments serve a common purpose, although at different stages in the development of a proposed project. The white paper is a preliminary technical proposal; it addresses a presumed need (or "opportunity"), which is some kind of knowledge that the customer might want to acquire. The point of the white paper is to describe a valid means of meeting that need. Later, as modified and developed after responses from the customer, the white paper becomes a section in the full proposal, which is the vendor's other formal transactional instrument. The proposal is a comprehensive claim statement; it contains a complete technical section, which is the bulk of the document, and it includes appropriate sections on research management and cost. Since the proposal is regarded as a proprietary document, a protocol is required in its forwarding from vendor to customer. The vendor must sign a memorandum of understanding (MOU) that gives the customer the right to read it.

Vendors write many more white papers than proposals, either as initiatives or in response to queries by customers. In either instance, white papers are often used by customers as source documents for developing BAAs and RFPs. Whether or not a prospective customer has indicated an interest in receiving it, the white paper offers the great advantage of informality. Because it does not legally require any response from the prospective customer, it avoids the expensive protocol that federal regulations demand in the treatment of proposals. From the vendor's perspective, this same convenience offers a different kind of advantage. Because its status is informal, the white paper can be modified and then recycled to different prospective customers. It gives the vendor more opportunities to sell.

Two other documents complete the transactional genres: the sources-sought announcement by the customer, and the sources-sought response by the vendor. Sources-sought is a brief description of a pending project and an invitation to bid on it; the announcement appears in the pages of the *Commerce Business Daily* (CBD). Published by the Government Printing Office, CBD is a daily list of federal procurement invitations, contracting awards, and other fiscal notices. Federal regulations require the publication of sources-sought prior to an RFP, and it usually appears a month or so in advance. The announcement briefly describes a pending project; it invites vendors to state their interest in bidding on it and to submit in writing their credentials for doing so. Interested bidders can then be screened, and some of them will be informed that their credentials do not fit the task. The sources-sought announcement serves several purposes. It saves cost by limiting the number of bids in the selection process; it also warns off ineligible bidders before they invest in making an expensive proposal. The document saves time and money for all parties. As part of a regulatory procedure, moreover, sources-sought provides an entry for new vendors into the market.

With reference to knowledge production, the more important document is the vendor's sources-sought response which is intended to represent the vendor's capability. The purpose of this document is to validate a right to bid on the pending project. Invited by the customer, it is candidly a set of claims, wherein the vending organization describes its research capabilities in terms of the knowledge product that a pending project is intended to yield. To all intents, the sources-sought response is an invited white paper.

These are the principal pre-contract genres: the customer's various contract vehicles, the RFP and BAA directed to vendors, the white paper and the proposal initiated by vendors, and the sources-sought exchange of invitation and response. They all reveal signs of positioning, of juxtaposing the customer's mission requirements and the vendor's research capabilities, as appropriate to the two different perspectives.

In addition to these seven transactional genres, customer and vendor both use one other instrument to contextualize their ideas: the formal conference paper, presented at joint meetings for specialists representing both customer and vendor within given fields of research. The meetings are usually sponsored by a defense agency or by a group of R&D activities. The technical papers that are presented or circulated help to clarify a customer's interests or to inform customers of a vendor's capabilities. The papers tend to define the state of the art in any subdiscipline or where subdisciplines from different areas merge. They emphasize needs or claims, depending on the author's perspective. And they engender follow-up discussions, both on and off the record.

THE TEXTS AND THEIR CONTEXT

In his review of genre scholarship in several different fields, Swales (1990) finds five criterial features that characterize documents as genres. All five of them confirm the generic nature of the R&D procurement documents. Two of these criterial features are obvious: the documents are classes of communicative events, and each class bears a functional name (or acronym). A third feature is an allowable latitude in form and style. Such latitude is evident in the texts of the customers' acquisition forecasts, which range from rather general to highly specific. It is also evident in the relative simplicity of the customer's sources-sought announcement as compared to the vendor's highly complex proposal. In addition, documents vary in their postures toward the audience. The vendor's proposal, which is categorically responsive to the reader's expectations, differs markedly from the customer's RFP, which has about as much sensitivity as a landlord's contract.

The two other features of genres are their shared communicative purpose, and the influence of such purpose on their form and content. These two features have the most rhetorical significance among the R&D documents. Swales puts foremost the feature of shared communicative purpose, and he spells out a tight conceptual relationship. Three characteristics denote this relationship: (a) each genre represents some common purpose, (b) each genre is the property of a discourse community, and (c) commonality of purpose links each genre to its discourse community. Such shared communicative purpose functions at different operational levels in the R&D discourse community. At a finite level, customers and vendors use the genres to serve their separate, individual purposes. At a more abstract level, all customers convey their needs through one subset of genres, and all vendors use another subset to claim their capabilities. At the most abstract level of operation, all customers and vendors mutually participate in the same genre system in order to expedite the production of knowledge.

But at all operational levels the notion of reciprocity is inherent in the R&D document exchange system. Engaging each other from two distinctly different rhetorical perspectives, customers and vendors successively reify some technological idea until they arrive at a viable research project. In the process, their rhetorical exchanges form a dialectic. In this reciprocity, the discourse community becomes a correlative of the marketplace. It functions through actions and reactions, and its genre system is social as well as linguistic.

In a social sense, the R&D discourse community is a community of shifting relationships. It is a construct of separate audiences that form and dissolve and form again in different alignments to serve the common purpose of defining an R&D project. In this respect, it has some familiar analogies. The legislative body is one; its members represent different political parties, yet engage in temporary, informal alliances that cut across party lines. All of the legislators are initiates, and the genres they use enable some kind of ruling communicative purpose to evolve through systemic abrasion. The judicial discourse community is another analogy. Bound by common rules of procedure, all lawyers utter typical, generic statements representing fixed perspectives, yet they can readily change roles in different litigations.

FORM AND FUNCTION IN THE DOCUMENTS

Shared communicative purpose is one of the two chief characteristics of the transactional genres. The other is the way in which their form and content reflect a communicative purpose. Form follows function. This feature clearly marks the R&D documents. They speak in a common language about the same things. The most obvious symptom of this fact is how they sound. Regardless of authorship, they speak in one voice and use the same language. The voice is public, authoritative, and impersonal; it is without modulation. Reading it aloud, one gets the sense of digitized sounds assembled in a sensible sequence. It all sounds robotic. Somewhere behind that voice there is authorship, but no author is present within the text; no persona claims ownership of the ideas that the voice presents.

The homogenous voice of the documents has some precedent in the ruling convention of technical communication in all fields. Therein, the writer is commonly assumed to be merely the purveyor of information that already exists; the author conventionally transmits, rather than creates, the substance of the text. The convention is reinforced in these documents by their corporate authorship. Whether written by one person or assembled by a group, the drafts of texts are commonly subject to sign-off procedures and supervisory amendments that favor boilerplate language. Electronic composing also plays a part in this de-authoring process. Like any new medium, it enhances conventional behavior. Composing on E-mail or on portable disks makes it easier for members of the authoring team to shape a text with impersonal amendments. The diction takes on a studied ambiguity, which is deemed by some to be useful.

Composing the text of a document is a collective enterprise, and it is usually an assembly process. Not surprisingly, therefore, the voice of the text is the voice of some corporate persona. It may actually reflect the voice of the person who drafted the text, as modified by those who edited it. Or it may reflect the presumed voice of the person for whom it was ghost-written. In any case, it is attributable to the person who signs off on it, or, more likely, to the office that releases it. The letterhead denotes the corporate persona.

Not only are the tone and diction the same from one document to another so is the form. In fact, the sound of distant impersonality is a kind of tip-off; it echoes a common architecture. In any carefully written discourse there is a tight relationship between a text's rationale and its structure. Form does follow function. In these careful documents, a common architecture is evident in what they speak of; they speak of problems. It is also evident in the stylized structures of information that convey the problems. The commonality of the transactional documents in this latter respect is striking, and it reveals a great deal about knowledge production—so much so, as to merit more discussion in the chapter that follows.

10

Formulating the Fundable Problem

The transactional documents used in R&D procurement have a characteristic architecture. Like the houses in a developer's plat, their texts are constructed along the same lines. Although they may have different facades, they all have the same interior space, following the same floor plan with a few optional variations. The textual space in these documents, by and large, is arranged according to one predominant rhetorical mode, one that necessitates the interpreting of information. Within the texts you can even see knowledge being produced. That process has our attention right now. How the architecture of these texts generates knowledge is our subject.

The context of these texts is already evident. Customers typically state their needs as problems, and vendors typically propose solutions. The documents that they exchange typically reference a particular subject and then explore its elements. As the previous chapter indicated, the sequence of the generic documents implies various pathways leading to a contract for an R&D project. Although most of the exchanges terminate before they arrive at a project design, they are nevertheless heading that way. Having seen that the documents move the procurement process along, we are ready to see how they do it.

The energy that moves the process is evident in the texts. As the document exchange proceeds, the texts of the documents focus on some problem or set of problems. Problem and solution is their principal mode. The texts may occasionally contain other rhetorical modes that serve an instrumental purpose in expounding on a problem; description, explanation, and analysis are appropriately supportive modes that appear from time to time. But narrative almost never appears; the texts have little or no story sense. Needs and claims are their substance: needs formulated by the customer in various problem statements, and claims formulated by vendors as solutions.

In a textual sense, however, the scenario is much more interesting than the simple repetition of action and reaction. The pattern is more involved, more elaborate. The argument of each generic text formulates a problem. In this respect, there is no distinction over whether the problem conveys a customer's need or a vendor's claim. No matter what the rhetorical purpose of any given text, its way of conveying that purpose is common to of the texts. They all formulate a problem, and in each of them the formulation is a means of knowledge production.

But a caution is appropriate. In conventional usage, a problem is seen to be a hardship of some sort, an adversity or an affliction that befalls a person or a group. Whether it is to be borne or overcome is another matter. The point is that it exists as an unresolved burden. But conventional usage is merely one way of comprehending a problem. There is a critical difference between perceiving that a problem already exists and deliberately constructing a problem for a purpose.

As a general rule, most situations can be formulated as problems, and usually for a purpose. The weather, one's diet, the national economy, or what is playing on television can all readily be converted into problems. The way one describes a situation can imply some problem, some unresolved burden, a fact that the news media characteristically exploit; problems make news. In this sense, problems do not just happen; they are constructed. Disasters and inconveniences happen, but problems can be made. They are constructs.

Problem formulation is a common mode of communication. A problem formulation conveys a sense of agenda, and it often anticipates some kind of response. In defense acquisition, the way problems are formulated determines the kinds of R&D that are ultimately undertaken. More precisely, the distinctive characteristic of problems in the R&D environment is the implicit premise that they will yield a solution, given enough time and the right resources. In fact, problems are formulated so as to make solutions inhere in them. In the idiom of R&D literature, they are fundable problems.

This chapter concerns fundable problems, especially the way in which they are formulated. It begins by describing the kind of problem that contains an inherent solution. Then it shows how this type of problem has been adapted to the formulation of needs and claims, the formulation used in the R&D procurement process. Finally, the chapter inspects the role of this problem type in a broader range of science and technology.

FORMULATING THE FUNDABLE PROBLEM

The fundable problem is a rhetorical construct whose purpose is persuasion. Its anatomy is derived from a familiar generic problem format that posits two elements: an initial condition and a goal, or desired condition. When they are juxtaposed, these two elements generate a third element that makes them interdependent. The third element is some way of reaching the desired condition from the present condition; in the language of problem theory, it is the "problem space" through which possible solution paths must proceed. This generic construct was defined by Allen Newell and Herbert Simon (1972) for the purpose of designing research in information processing.

The act of juxtaposing those first two elements, an initial condition and a goal, makes the third element an imperative. The imperious third element, a solution, is exactly the point of the information processing problem; the whole construct calls for a solution. Suppose that you have scheduled a trip to see an important client tomorrow morning. You are downtown at 5:00 p.m. and you want to board a plane in an hour at the city's airport twelve miles away. The discrepancy between being downtown and arriving at the airport on time immediately engages you in deciding how to get there. By its nature, the problem construct captures your attention and your commitment to solving it. You confront it.

Even as you consider your alternative routes to the airport, you are aware of the constraints in your initial condition that may impede your achieving your goal. You are downtown at the rush hour, your car is in a parking lot four blocks from your office, the expressway is under repair, and it is raining. The constraints all heighten the urgency of your commitment to weigh the risks of the alternative routes. Which one shall you take? The repair on the expressway may tie up traffic. An alternate route with less traffic is twice as long. There is a radical shortcut along a dirt road that might be impassible in the rain. As you weigh these alternatives, you are already testing the problem space.

You begin to solve the problem by refining it, perhaps by modifying one of its elements. You cannot modify your present condition, but if all the solution paths are hopeless, you can modify your goal. Might you exchange your present flight for a later departure and then choose the longer route to the airport? If not, does your goal need to be more radically modified? Might you telephone your client and change the appointment? If so, can you somehow work the delay to your client's advantage—a late lunch, perhaps? Or, if the delay is longer, a new database that you can prepare in the meantime? At this point your dialogue with yourself, and perhaps with your client, is beginning to evolve a different problem than the one that confronted you only moments ago; the formulation is dynamic. Although there is still the awkward situation of how to get to your car in the rain, that is now a separate problem; it can wait until you have decided how you are going to deal with your client. My point in invoking this episode is to dramatize the idea that there is a solution lurking somewhere in the construct, if only you can find it.

Getting to the airport on time necessarily involves you in reformulating the original discrepancy between an initial condition and a desired condition. Therein lies the analogy to the R&D procurement process. Solving the generic problem entails reformulating it—interpreting the available information, and massaging it into some other construct. The key to reformulating the problem is to factor it into subproblems. "Divide and conquer" is the rule in information processing that software programmers abide by. The rule is a thumbnail version of the model for formulating problems in engineering design that Vincenti (1990) describes. His model of successive activity levels is a construct for factoring a large-scale design problem into smaller, discrete problems that are successively nested in one another.

Factoring a problem presents solution paths that are progressively more finite, more readily dealt with. It also yields new relationships among the variables of the original problem, resulting in new possibilities and new combinations of information. Factoring a problem is generative; it raises the odds on producing new knowledge.

PROBLEM FORMULATION IN NEEDS STATEMENTS

The factoring of problems is characteristic of customers' needs statements. The military departments publish annual forecasts of their technology needs. Each publication contains a set of Broad Area Announcements (BAAs) that cover possibly 50 areas of technology development for which funding is available. Each BAA includes a broad range of needs, yet the statements must be detailed enough to elicit responsible inquiries from vendors. The texts must pack a lot of information in a small space and make it intelligible. Given these requirements, BAAs tend to be constructed as problem statements in a way that takes advantage of factoring.

The texts of such BAAs typically move their subjects down the ladder of abstraction. They factor high-level problems into low-level problems; they convert perceived needs into action-oriented algorithms. *Guide to Naval Aircraft*, an acquisition statement issued by the Naval Air Propulsion Center (1991), is a good example of this step-down sequence. A needs statement comes first, then a set of goals to meet those needs, followed by a description of the level of effort anticipated for the research. This BAA pertains to the technical field of advanced propulsion for turbojet and turbofan engines. Beginning with an abstract statement of need, the text presents successively reduced levels of abstraction.

The needs statement begins with an open-ended goal, calling for "substantial improvement" in two critical measures of performance.

> Aircraft operational needs for extending radius, time-on-station, payload and penetration speeds require a substantial improvement in engine thrust-to-weight ratio and specific fuel consumption (p. 1f).

Substantial improvement is then factored into significant advances in propulsion component and subsystems technology.

> Development shall focus on key Navy air-breathing propulsion component and subsystem technology where significant advances would have a significant impact on Naval aircraft mission performance, operational capability and affordability.

The focus on components and subsystems is then specified:

> Developments in fan, compressor, combustor, high and low pressure turbines, augmenters, exhaust nozzles, mechanical systems, and opto-electronic controls technology.

And the component specifications include relevant research:

> Propulsion integration studies of innovative concepts, advanced components and systems are contemplated.

Finally, a section that specifies Level of Effort narrows the needed research to component design and performance verification tests:

> Offerors shall consider three to five year development projects consisting of detailed design, fabrication of demonstration engine quality hardware, and design and performance verification (analytical and test) phases.

At this lowest level of abstraction, the acquisition forecast concludes. It leaves to prospective vendors the next step in formulating more detailed problems, which will also doubtless be factored in the project proposals.

Factoring a problem into constituent parts is commonplace in this procurement literature. It shows up even within paragraphs in *Long Range*

Acquisition Estimates, FY91 and FY92, issued by the Aeronautical Systems Division of the Air Force (1990). Consider, for example, item ASD-91-32, a two-paragraph sequence about technology integration for unmanned air vehicles (UAVs); it typifies the prevailing syntax in this volume of acquisition estimates. Each paragraph in the sequence follows a step-down series of successively lower levels of abstraction. Each proceeds from global need to local need to specific elements of the local need. Parsing the dense sentences is rather like picking hair thorns from a cactus out of one's finger, but the analysis is rewarding. The two paragraphs present a compact problem formulation that anticipates, and even contains, a solution. They say, in effect, "Tell us what we need to know in order to commit serious money to preliminary planning for UAVs."

The step-down sequence is a series of five "needs" conflated into two paragraphs, one labeled "Objectives" and the other "Technical Approach." Using syllogistic logic, the "Objectives" paragraph telescopes two global needs into a local need:

- The need to reshape the military force structure after 1989 to better adapt to the new world condition,
- The need to use force with the minimum of exposure to service personnel, and [therefore]
- The need to use UAVs as a complement to current force structure in the face of a complex threat environment.

In order to meet the (local) need for the use of UAVs, two other needs are stated in the second paragraph, labeled "Technical Approach." Each of them is a need to "determine" certain kinds of information. (Determining information, in this case, entails gathering, collating, and assessing it.) The first need in the Technological Approach is to determine three kinds of information:

- Opportunities for operational employment of UAVs
- Force integration requirements, and
- The magnitude of technology investments required.

The output of these tasks generates a final need, which is to determine more specific information that anticipates engineering design:

- The need to determine operational requirements, performance requirements, and technology requirements in order to develop conceptual designs of UAV usage.

The statement concludes by anticipating that the Aeronautical Systems Division will use the proposed designs for possible modification of the present force of manned aircraft.

Several textual features of this statement merit special attention. The first is the factoring of a problem into subproblems, which are more precise needs. The second is the increasingly concrete nature of the formulation, wherein the first three needs progressively define a problem that the last two needs begin to solve. The third, and probably the most significant, feature is the generative nature of the statement.

The statement is generative in that its indication of a research design, which concludes the statement, is an interpretation of the information that has preceded it. The indication of a research design, however tentative,is new knowledge. The customer still does not know exactly what information it needs to proceed with R&D for the UAVs; finding that out will be the vendor's task. But the customer does know the *kinds* of information it needs. Each kind is like an arrow pointing the way to a solution of some part of the problem. The problem formulation is an algorithm that produces knowledge even in the process of calling for knowledge production. It has what literary critics would call organic form. It does what it says.

THE TEXTUAL FUNNEL

The Air Force's solicitation of research on UAVs and the Navy's solicitation of R&D on aircraft propulsion components are two versions of the fundable problem. Each one presents a construct of the same three elements—a problem space interpolated between two different conditions. This is the seminal construct from which a solution can grow. Then each solicitation factors the construct into subconstructs of the same three elements, but at successively lower levels of abstraction. And the factoring gradually formulates new knowledge. By invoking more and more concrete information to specify the original components of the problem, the factoring thereby anticipates the parameters for a solution. Each of these BAAs represents the first step in the design of an R&D project: a forecast of the shape it will take. The intended project, if it happens, will yield still further, more specific knowledge as a result of the project design. The customer's formulation of the problem is directive; it is a vector of information.

The same formulation that is used by customers to state their needs is also used by vendors to state their claims. No matter what the communicative purpose of a statement may be, the same kind of problem formulation will serve the purpose. In either case, this formulation produces knowledge; it also channels the new knowledge in a way that serves the communicative purpose of the author.

An analogy may help clarify this textual phenomenon. The structure of a fundable problem can be caricatured as a funnel; it progressively channels predications into a directed stream. At the upper end of the funnel's cone, the formulation begins by posing some large discrepancy between the present condition and the desired condition. To recall the term used by Newell and Simon, it poses a "problem space" of global dimensions, to which the reader can readily agree. It then narrows this discrepancy by naming a relevant subgoal and the corresponding constraints that will impede the achievement of that subgoal. This lesser discrepancy, at the lower end of the cone, can then be addressed in terms of the capabilities needed to resolve it. Finally, down the stem of the funnel, the narrowest and longest part of its structure, the statement describes the capabilities that will be needed to resolve the lesser discrepancy.

In the exchange of documents between customer and vendor the textual funnel works equally well either way; it serves the customer, and it serves the vendor. The customer uses the textual funnel to provide a rationale for specifying the kind of capabilities the vendor will need in solving the problem. Similarly, the vendor arranges the customer's specifications in the form of another textual funnel to provide a rationale for declaring its own capabilities.

The figure of the funnel is so clarifying that it might seem to represent this step-down formulation of a problem as an autonomous entity. So a note of caution is appropriate. This type of problem formulation is a function of its social context. It is important to bear in mind the relationship between the formulation and the people who use it. Through frequent usage over time in the R&D community, the fundable problem has become generic. In this respect, it has a formative influence on any person who uses it in composing an argument. Alastair Fowler (1982) points out that this influence is a property of all genres. Every genre offers some configuration of structures and procedures, some set of formal ideas by which to order the information that one wants to convey. This is true of the fundable problem. But the formal ideas that it presents are options, not mandates. The textual funnel is an adaptable tool; its use can be modified to fit different situations.

As I suggest, vendors also use this generic problem formulation. Their versions of it are embedded in the texts of their white papers and their proposals. Vendors can funnel information in this fashion whether they are responding to a BAA describing general needs or to an RFP for a specific project. Either way, the funnel accommodates the customer's expectations. If vendors are responding categorically to a customer's BAA or RFP, they can be sure that the solicitation has already defined some part of the problem. It has already shaped the cone of the funnel and arrived at a statement of the kind of solution that is needed. If vendors are initiating an offer, on the other hand, their proposals will need to start at the top end of the cone, delineating the more global problem first and then progressively refining it. In either case, however, the vendor concentrates on the part of the problem marked by the stem of the funnel, that is, on the narrower statements of the capabilities it will commit to the proposed project.

PROBLEM FORMULATION IN CLAIM STATEMENTS

Capability is what the customer buys and what the research vendor sells. So it is not surprising that the vendor would adapt the generic problem formulation for the purpose of persuasion. The vendor uses the formulation in a way that features the capability it can bring to bear on the customer's problem. Validating one's claim of capability entails offering two kinds of warrant. One is demonstrating a thorough understanding of the problem; the other is specifying the details of the particular solution path that the vendor is offering.

Several brief examples help to clarify how the funnel formulation works to the vendor's purpose. To begin with, capability is a high-level abstraction; merely claiming it carries no credibility. It must be demonstrated. But the task for which the capability must be demonstrated is somewhere in the future. So the vendor's capability must be claimed on the basis of something else, something already achieved. Here is where the formulation of the fundable problem serves the vendor's rhetorical purpose. It enables the vendor to make many smaller but related claims that can be supported by evidence. Factoring a problem into subproblems readily accommodates this tactical effort; it multiplies the vendor's opportunities to claim and, if possible, demonstrate a number of finite capabilities. Each subproblem provides an opportunity for interpolating a claim for some constituent capability that will be needed in the future task for which the vendor is bidding.

The following skeletal versions of two technical proposals readily illustrate the factoring process at work for the vendor. The texts of both proposals lead the reader step-by-step down the ladder of abstraction. It takes only a few predications to show how a project director can break down a problem and establish a new vector that features the vendor's capability.

What follows is my paraphrase of a few excerpts from the first proposal. The excerpts are indented. My interpolated comments about the excerpts begin at the left margin. Thus, there are two voices talking here, the proposal's voice and my own. The first proposal is about the fabrication of composite materials.

Global Problem Space. In the coming generations aerospace structural applications will be subjected to extremes in loading and temperature far exceeding the capacity of present-day structures. Such materials are likely to be ordered compounds, composites with ordered compound matrices (and fibers), and nanoscale equilibrium structures.

Local Problem Space. There is a need to understand the deformation mechanisms, constitutive behavior, and damage accumulation in these future materials. But such stress states have been relatively unstudied:

- The materials are difficult to produce, and they are in short supply;
- There is no clear "best choice" among these materials; and
- The deformation mechanisms and constitutive behavior of these materials are very different from what occurs in traditional materials.

These conditions present a number of nontrivial problems.

Solution Strategy. To develop a general framework for understanding the behavior of these materials, the vendor proposes to test one ordered monolithic compound and a related composite under complex conditions of load and temperature, and to develop reliable three dimensional information on damage accumulation, using modern techniques.

Proposed Tasks

- Testing two specific composite materials that the laboratory already has available,
- Explaining the theoretical basis for the testing of these materials, and
- Explaining the replicable procedures for testing them.

The Vendor's Capabilities

- The vendor's experience with composite materials.
- The modern techniques in the vendor's laboratory that will make the proposed procedures possible.

The strategy of this proposal is evident from the fact that the vendor already has two composite materials on hand, available for testing. What the vendor needs is a credible reason for the sponsor to fund the cost of testing them. In this circumstance, the step-down formulation of a problem serves the purpose; the funnel model is functional. The projected need for such advanced materials (at the top of the funnel) provides a rationale for proposing a solution (at the bottom of the funnel's cup). Then the solution leads (down the stem of the funnel) to the vendor's capabilities for testing these materials. Here is a common instance of a fundable problem. The sponsor's presumed need and the vendor's declared capabilities are all of a piece. The vendor is telling the sponsor, "What you really want is what we know how to do."

The text of the other proposal reveals a variation of the basic formulation. It breaks down the problem twice, first to opt for an innovative solution and again to offer a procedure for testing that solution. As with the first proposal, I have paraphrased a few excerpts from the text. The proposal concerns countermeasures against guided missiles that seek and destroy aircraft. Its formulation begins in the familiar way.

Global Problem Space. Current guided missiles resist present countermeasures (such as jammers and flare ejection circuits) that cannot now be tailored to individual missiles. A solution is needed.

Anticipating a Local Problem Space. The ideal solution of mounting a directed-energy weapon on the aircraft is impossible because of the weapon's excessive weight.

Rationale for a Local Solution. An alternative approach would be to illuminate the threat with sufficient power to cause low-level damage, but to pulse the energy in such a way as to disrupt the missile's target seeking system.

Local Solution. The proposed work is to determine the usefulness of this alternative approach by using digital models, which this laboratory has specialized in developing and using, in order to recommend possible modifications to present electronic systems.

Here is an ingenious instance of the funnel formulation. The proposal waives the global problem, which is beyond the vendor's capability; it substitutes instead a local problem which the vendor does have the capability to work on. By changing the goal, the author reduces the problem space to something manageable.

Each of these proposals factors the global problem in a way that directs attention to a local instance of the vendor's R&D capabilities. The first proposal, about compound materials, guides the customer to the vendor's inventory of materials and to the vendor's procedures and equipment for testing them. The second proposal, about guided missile countermeasures, leads the customer to the vendor's extant computer models of variations in energy pulsing. In both cases the problem formulation provides a rationale for building the vendor's capability claims into the task analysis.

The exchange of problem formulations between customer and vendor produces knowledge in two ways: within each text, and in the exchange of documents. Within each text, the problem formulation progressively narrows the problem space in such a way as to incubate a solution; viable or not, the solution is new knowledge. In the exchange of documents between customer and vendor, the sequence of iterations from one perspective to another provokes new realizations or qualifications; these, too, are new knowledge. In both cases, there is a generative relationship between the given and the new. The given information is interpreted from a different perspective, thereby producing new knowledge.

This compounding of knowledge is fascinating to watch. Reading a sequence of problem formulations written from two perspectives is rather like reading a detective story. However improbable that might sound, there is an analogy here to the quintessential mystery, the puzzle of the sealed chamber. It begins with an improbable situation for which there is no evident explanation. The puzzle is then revisited, and each visit yields something new, some increment of a solution that turns out to be not only possible but convincing. In formulating and reformulating the puzzle by stages or degrees, the detective discovers the solution. The problem was the genesis, and the solution is its genetic derivative.

PROBLEM THEORY AND THE RANGE OF PROBLEMS

The fundable problem is a form of argument. It is a structure for converting information to a different order of value so as to serve some particular rhetorical purpose. It is a highly developed version of the information-processing problem delineated by Newell and Simon (1972) for problem solving in the information sciences. The commonality of the information-processing problem in the defense R&D community is not surprising, since the defense critical technologies are so information-oriented. Because it is such a familiar research tool, its prevalence in the field of information technologies is also not surprising.

In the broader context of problem theory, however, the information-processing problem is only one type of problem formulation. It is conceptually different, for example, from the problem construct commonly associated with scientific inquiry. Although both types of problem produce knowledge, they do so in different ways and for different reasons.

The scientific problem, as described by Stephen Toulmin (1972), has its basis in philosophy. It is seen to be a discrepancy between what we want to explain and what we are able to explain. "Problems arise where our ideas are at variance either with Nature or with each other," Toulmin says. A problem in this sense is a discrepancy that provokes a crisis in explanation. That discrepancy drives us to improve our perceptions of the natural world "by identifying problem areas in which something can be done to lessen the gap between the capacities of our current concepts and our reasonable intellectual ideals" (p. 149).

The information-processing problem and the problem that Toulmin described are superficially similar. Both problem types begin with a discrepancy between an initial condition and a desired condition, and both constructs seek a resolution of the discrepancy. But the similarity ends there. The two problem types are conceptually different, and the discrepancies that they posit are of different orders. Because the two problem types serve different purposes, they also serve different intellectual functions. The information-processing problem is constructed to provoke a pragmatic solution that gets rid of the discrepancy. The problem of scientific inquiry that Toulmin describes exists to develop better ways of explaining the discrepancy. In this respect, curiously, the purpose of the problem that Toulmin describes is much like the purpose of problem constructs in the humanities. (In the extreme, such constructs reify the irresolution of discrepancies.)

Reviewing the literature of problem theory, Michael Carter (1988) perceives these two problem types to be polar. To explain their polarity he invokes Simon's notion of well-defined problems and ill-defined problems. The distinction pertains to the relative force of the constraints that any given problem imposes on a solution; the well-defined problem incorporates more constraints, and the ill-defined problem incorporates fewer. At one extreme, the well-defined problem is a crossword puzzle or a game of solitaire; at the other extreme, an ill-defined problem is a complex of incongruities that arise from different perspectives and different disciplines, thereby complicating any solution. Using this polarity, and other precedents in problem theory, Carter proposes a model to make it easier to recognize various problem types. The model is a continuum that stretches between two extremes, the well-defined problem and the ill-defined problem. All other problem constructs in this model are positioned along the continuum according to the relative number of constraints they incorporate.

Carter's purpose is to show teachers of composition that problems are not necessarily static; as a writer composes a problem, it tends to move along the continuum, to become more well defined. His analysis aptly fits the life and career of a fundable problem. With one important difference, the fundable problem is essentially a well-defined problem; the two differ only in purpose. One is an analytical tool, and the other is a rhetorical tool; the fundable problem is deliberately intended to persuade. But otherwise it fits Carter's model of a problem moving along a continuum toward more precise definition. Its recursive authorship makes it dynamic. In the defense R&D procurement process, a fundable problem does not usually exist by itself; it is part of a reciprocal exchange of problems, part of the social context. In the exchange of documents peculiar to this process, the formulation of a problem is derived from previous formulations. The fundable problem does not remain at any one location along the continuum. It evolves.

To be fully understood, the fundable problem must be seen as a process. It unfolds; in the course of its formulation the constraints change to accommodate a solution that becomes part of the next problem. This discourse system is intertextual; text and context merge. Within the system, each problem has a history—and, possibly, a future. As it reformulates the information it inherits, each problem produces new knowledge. In this reformulation, process is all.

11

Capability Statements: The Truth But Not the Whole Truth

The most prodigious genre of documents in the defense R&D community is the capability statement, which is an organization's advertisement of itself. Capability statements serve a function in the sale and purchase of R&D. Like the transactional documents that lead to an R&D contract, they have their own distinguishing features. Chief among these features is a communicative purpose that is understood by all members of the discourse community. We turn now to that purpose and to the various means of implementing it, as we explore the knowledge-producing properties of this genre.

The rationale of the capability statement in the defense R&D market lies in the uncertain nature of the product. No matter what the scale of endeavor, the output of R&D is obviously unknown at the outset; the uncertainty exists equally in an individual R&D project or a whole program of projects. What a customer buys is the vendor's presumed capability of completing the intended work satisfactorily. And what the vendor sells is essentially a promissory note based on its claim of capability. In this scheme of things, if the satisfactory output of R&D is the whole truth, the promise is only a part of it. Yet it is a viable part.

Claiming capability is essentially a vendor's undertaking. In the dynamics of this market, however, the customer also assumes the role of vendor. The customer needs to represent its achievements favorably to those who have funded them. This need occurs all the way up the line of command, from individual R&D activities to the DoD itself. In the most conspicuous sense, DoD must convince the Congress and the public of the viability of its management of defense technology development. At some point, therefore, each customer in the line of command must assume the role of vendor; it appropriates to its own marketing purpose the aggregate capabilities it has earned through its previous initiatives. This fact accounts for the enormous quantity of capability statements that circulate within and beyond the defense R&D community. Every organization, vendor and customer alike, is obligated to explain itself and to do so continually. So it issues capability statements.

The breadth of the genre is bewildering. On one fringe of it is the official chronology meant to document an organization's importance for the record. Such is the *40th Anniversary of the Joint Services Electronics Program* (JSEP, 1885). On another fringe are the public position papers that describe policy, although pointing with pride, in the process, to organizational achievements that make the policy viable. The *National Security Strategy of Engagement and Enlargement*, issued by the White House (1994), is one prominent example. Others are the statements before congressional subcommittees by officials in the defense establishment, which are then circulated. Declaring an organization's capability is commonplace.

Notwithstanding the breadth of the genre, however, it can be understood through a relatively few features common to all capability statements. Their main feature, their communicative purpose, is also their rationale. The capability statement exists to help position an organization advantageously with the public or within a market. Positioning is its purpose. Another common feature is a textual structure that accentuates this purpose; in a typical capability statement, this structure is evident in the sequence of a few textual elements. After first noting the rationale of the capability statement, this chapter explores the architecture of the genre and how it generates knowledge.

THE RATIONALE OF THE CAPABILITY STATEMENT

Although the output of R&D is knowledge, the product for sale in this particular market is the capability of producing that knowledge. As we know, the vendor's capability contributes materially to the design of the R&D project. Then that project becomes part of the vendor's new knowledge base. But there is more to this knowledge base than we have so far explored; there is a significant difference between the output and the outcome of an R&D project. The output consists of the deliverables that were agreed to by contract. They constitute one kind of knowledge that the vendor produces. But the outcome of the project is broader. It includes a different kind of knowledge, which the vendor acquires in the process, a kind that is nonquantifiable and not precisely measurable. It is the knowledge of how to produce knowledge.

This acquired knowledge is a capability. As Machlup (1980) explains, it consists of knowing *how* as distinct from knowing *what*. It has been earned through the experience of carrying out the tasks specified in the Statement of Work (SOW) that delineated the R&D project. Simply by experiencing the process of completing the tasks, the R&D team learns new things about a given subject. It learns about the relationships of certain variables, the behaviors of component parts, the viable measures of reliability, and a host of other perceptions.

Although these perceptions may or may not have been logged and reported at the time, they can later be generalized to apply to other tasks in other contexts. Such bits of knowledge do not pertain exclusively to the project that bred them; they belong rather to the domains of the science and engineering disciplines that the project entailed. The vendor attempts to generalize such knowledge in a way that will accommodate new circumstances; the purpose is to provide the basis for other claims of capability. For the customer, this knowledge is also an achieved value; it represents an investment in the vendor that may pay off in subsequent projects. The vendor's newly earned capability will support the customer's mission objectives through later R&D initiatives.

As we saw earlier, to market a product is to position it advantageously from a customer's point of view. Once a prospective customer has been identified, marketing coordinates all of the operations needed to shape the product to the customer's needs. In the commercial market, these operations include product design, production and pricing, packaging, advertising and promotion, and distribution. All of them depend on positioning the product with reference to the customer's needs; they depend on creating an image for the product that relates it to those needs. In short, marketing a product creates a context for selling it, and positioning the product is central to marketing it. This axiom applies to all markets, including the market for defense R&D.

Although advertising is only one of the instrumental operations in marketing, it is the one that the customer usually sees first. It is the most conspicuous means of relating the product to the customer's needs. Advertising confirms the positioning of the product; its task is to convey that positioning to the customer's mind, to translate it to the customer's perspective. In this particular market, the task is performed by the capability statement. Although the customer would like to buy the output of R&D, that is impossible, since the output is unknown until the project is complete. So the only product in this market is the vendor's *presumed* capability. It is really only a virtual product, and the customer's dilemma is to have to choose among many presumed capabilities. The task of each vendor's capability statement, therefore, is to convert a presumption into the conviction that this particular vendor can perform the R&D that the customer needs.

Selling the product in this market, as in most other markets, entails advertising. To forego advertising is not an option, since advertising reinforces the product's positioning. But there is distinction between this virtual product and the more typical commercial products in other markets, that puts a different stress on positioning. For the R&D vendor, every new project represents a potentially new market. Even though it may be an easier market to penetrate if the customer remains the same, the competitors will be different. Even more important, the SOW will be new each time. Since the defense S&T market churns with new initiatives, the vendor must always engage in repositioning. To be sure, the R&D vendor tries to acquire and maintain a corporate image that assures quality and dependability, no matter what the new project might be. But a reputation for quality and dependability is not by itself enough; it must somehow be made relevant to each new initiative.

The burden on the capability statement, therefore, is to keep the vendor's virtual product—capability—poised to accommodate each new initiative. Although the fact may be easily overlooked, advertising is not a force by itself; it exists to reinforce the product's positioning. The positioning in this case entails a vision of inertial success. So the capability statement must convey a vision of future attainment based on past achievements, and it must marry these two concepts in the customer's mind. Getting the customer to agree to that marriage is a delicate task, but it is also commonplace, as anyone applying for a new job would probably recognize.

THE ESSENTIAL CAPABILITY STATEMENT

The job candidate's basic tool is the personal capability statement, better known as a professional resume or a *curriculum vitae*. The textual elements in the basic corporate capability statement are strikingly similar to those in a job candidate's professional resume. Because the latter document is so familiar, analyzing it provides a shortcut to the corporate document.

The personal resume is essentially a set of lists that pertain successively to one's professional credentials: previous employment positions, formal education, honors and awards, special interests, and the recommendations of persons with authority. Preceding the lists is a brief indication of the kind of position that the candidate seeks; it is usually called a career objective. The skeletal text of a resume is a set of slots for holding successive pieces of information; the slots have conventional labels (see Table 11.1).

The conveyance of one's capability depends on how these bones are arranged and then fleshed out. To begin with, the career objective can incorporate any features of the desired position that the job candidate wishes; in any case, it represents a vision of some sort. Thereafter, certain conventions allow for arranging the information that follows, so as to create emphasis where it will be most useful. For easy reading, the conventional sequence follows the logic of a news story, with which all readers are presumed to be familiar. It presents the most newsworthy statement first, followed by successive items of relevant information in a decreasing order of urgency. Following this convention, the format inverts the chronological order of the listed items so that the most recent item appears first. By the same logic, the sequence of the lists may vary, depending on what the candidate wants to emphasize.

When this skeletal structure is fleshed out, it conveys a narrative sense. To be sure, it is a singular kind of narrative; it begins in the future and travels backward through the present to the past. Narratives, of course, can go backwards or forwards, and some narrative structures accommodate both directions. The flashback in the chronology of a novel, for example, is a common fictional convention; it is a means of using past events to explain the present. The analogy to fiction is fitting. In a capability statement, the inverse narrative relates a presumed capability in the future to a set of present and past performances. It presents a credible fiction.

TABLE 11.1

J. P. Doe
(Address and Telephone Number)

Career Objective	A Statement Conveying Future Vision
Employment Record	List
Education	List
Honors & Awards	List
Special Interests	List
References	List

The coherence of the text becomes apparent when the slots that represent each category are filled with appropriate information. The prefatory statement of the personal resume is the candidate's vision of the future, and the listed slots contain information used as evidence to support it; notably, each slot contains an achievement (which is usually highlighted in some way). The top-down sequence is a convenient way of linking the vision to the evidence (see Table 11.2).

TABLE 11.2

J. P. Doe
(Address and Telephone Number)

Career Objectives	Senior management in a manufacturing company with international markets
Employment	
1990–1996	Fabric Cutting, Inc. Vice-President • Instituted computer-automated fabric inspection system. • Developed internal personnel promotion.
1987–1989	Textile Printing and Dyeing Co. • Director of Training, TPD, Costa Rica • Instituted workforce mentor program
1983–1986	Office Supplies Assistant Head, Marketing Department • Drafted comprehensive plan to discontinue low-margin product lines
1979–1983	Glassware, Inc. Assistant Superintendent for Southeast • Designed cost-cutting procedures
Education	
1982	• MBA degree, XYZ University
1978	• BS degree, Industrial Engineering, ABC College
Honors & Awards	
1992	• Highest Merit Award, *Textile Journal*
1981–1982	• PDQ graduate fellowship
Special Interests	Spanish language Hang Gliding
References	[Several names and addresses]

From this cluster of lists, a narrative flows forth. How do we account for the transformation? Actually, it occurs in the reader's head, not on the page. Influenced by convention, by a deep habit of converting top-down information into narrative form, the reader supplies the continuity. That continuity facilitates the task of the capability statement, which is to get the reader to yoke past achievements to future capability. The text merely supplies the clues in an easy order for the reader to follow.

Except for their labels, the categories and the logic of the sequence in which items occur are virtually the same in the personal resume as in the corporate capability statement. One set of corporate capability statements from the Georgia Tech Research Institute (GTRI) illustrates this simple textual structure: a top-down sequence that begins with a vision statement followed by listed achievements. GTRI is an R&D shop with nine discrete laboratories. With a gross income of about $100 million a year, it is one of DoD's top 100 university-based suppliers of R&D. To respond to routine inquiries and provide an in-house reference for writing proposals, GTRI maintains an inventory of capability statements; they roughly correspond to the special fields of its constituent laboratories. All of the statements have the same format: two sides of a single file-size sheet with a three-column text, illustrated by several figures or photographs. They all begin with a vision statement, followed by achievements and resources, that are listed in varying order, depending on the respective strengths of the different laboratories. These documents are generic.

Each of the vision statements conveys a claim. One statement, featuring radar capabilities, claims GTRI's national reputation for excellence and comprehensiveness in this broad field of R&D. Another, featuring engineering sciences, candidly invites opportunities to apply its hardware, software, and systems-design capabilities. Still another, featuring computer science and information technology, begins with a description of its mission. The capability statement that features energy and environmental studies begins by noting a national crisis, which GTRI is already addressing. The statement of manufacturing technology capabilities begins by admitting that the reason for its success is its responsiveness to the needs and schedules of its sponsors. Some of the claims differ notably from others. The Concept Analysis Laboratory, for example, emphasizes its focus on getting operationally relevant results. The Environmental Science and Technology Laboratory, on the other hand, features a broad spectrum of issues addressed by multidisciplinary teams.

The claim in each case is followed by lists of categorized information that pertain to the claim, and each statement concludes with a name and telephone number to call for further information. As for the categorized information, each statement claims four or five kinds of capabilities. The most frequently cited are:

- Key areas of expertise
- Component disciplines
- Integration of disciplines
- Experience (variously described)
- Typical programs
- Research facilities and equipment
- Key defense or industrial sponsors
- Relationships to national priorities
- Outreach training and technology transfer

Although the listed items under these headings are neatly separated, the information is densely packed, and the language is technical. These statements assume a reader's working knowledge of the subject; they are addressed primarily to insiders.

SOME UPSCALE MODELS OF THE GENRE

I mentioned earlier my arbitrary classification of documents into three tiers according to their intended audiences; tier-1 audiences are the most generalized, and tier 3 the most specialized. According to this ranking, the GTRI capability statements are tier-3 documents; they are addressed to specialists who already know the context of the subject. Nevertheless, their top-down structure is equally applicable to tier-1 documents written to general public audiences. Such audiences need much more context in order to understand the claims that are being made. Tier-1 statements are the documents of choice among the RDT&E activities within DoD. One continual priority of these activities is to persuade many different lay audiences of their importance, their excellent performance, and their cost-effectiveness. So they assume the role of vendor and appropriate the capability statement to help position themselves advantageously.

The tier-1 capability statements tend toward the deluxe. Unlike the low-budget transactional documents that pertain to R&D projects, these are neon statements. Typically printed in four colors on glossy stock and lavished with high-resolution photographs, they range in length from 12 pages to 100 or more. They look much like a large corporation's annual report; designed and printed professionally, they convey the impression that production cost was no obstacle. The typical structure of these documents replicates the structure of the personal resume: a vision statement followed by categories of information presented as evidence to support it. Bells and whistles are often added, including logos, slogans, and keynote greetings from the commanding officer, accompanied by the officer's photograph. Yet the textual structure remains the same.

Human Systems Center Products and Progress (1993) is one example of this upscale model. It is a book, 143 pages long. The first four pages present the vision in various formats, and the next 139 pages present evidence to support it. The vision evolves one page at a time. "There Are No Unmanned Systems," we are told on the title page. The next page lists the values of the organization, headlined as Vision, Missions, Goals, and Guiding Principle. The guiding principle is alleged to make the vision come true: "To make the human the heart of aerospace systems and operations" (p. 2). On page three, the commanding officer of the Air Force's Human Systems Center (HSC) candidly asks, "How well do we meet your human systems needs?" The next page is a tear-out survey that invites the reader to judge this publication and to ask for more information about any of the HSC technology areas. There is also space for answering the question, How can HSC better serve your needs? The reader is invited to return the survey form to the commander's office or to the HSC marketing office at Brooks Air Force Base, Texas.

Page five of the capability statement forecasts the kind of evidence that will follow in support HSC's claim of dedication to the human individual in the technological system. A graphic format of HSC's tactical organization is presented in a four-color diagram of the Technical Planning Integrated Product Teams that constitute the operations at HSC. What follows is a series of one-page layouts, with photographs, to describe 121 R&D or T&E programs arranged in 17 subcategories listed under Crew Systems, Human Resources, Aerospace Medicine, and Occupational and Environmental Health. The program layouts form the core of the book. They correspond to the categorical achievements in the job seeker's resume. As for the usual references at the end of the personal resume, the book lists 45 names and telephone numbers to call to learn more about HSC's programs.

A shorter but equally lavish capability statement is *Science and Technology 1994*, a richly illustrated, 28-page booklet issued by the Naval Undersea Warfare Center (NUWC, 1994), Newport Division. The Center's claim to excellence inheres in the theme of the changing world threat to national defense, now that the Cold War is over. The gist of the vision is the Center's confident response to the new threat environment; NUWC is ready.

The singular quality of this text is the deliberate care with which the vision is developed. It is formulated incrementally by various repetitions that extend to a quarter of its length. First, a salutation from the Center's technical director emphasizes the challenge posed by the changing military threat. Next, a citation from the congressional OTA describes the appropriate functions of defense S&T laboratories. Then a headline, "The Continuing need for Technological Advancement," anticipates several paragraphs of warning: regional frictions in the post-Cold War world pose the danger of exposure to Soviet weapons in the hands of rogue nations. Next follows DoD's rationale of the need for superior technology as a force multiplier. Finally, the reader is led to the challenge facing naval technology: to change from open-ocean fighting areas to shallow and littoral waters, and to develop the enabling R&D during a period of shrinking defense budgets. It is a significant challenge. If the reader should happen to be a congressional staffer, so much the better.

Following the generic structure of the capability statement, successive categories of information are cited to support the vision. They include:

- physical plant
- personnel resources
- the S&T program
- Transitions/demonstrations
- defense technology conversion

The narrative clues are highly condensed, with about two-thirds of every page devoted to photographs and colorful graphic figures. Along the way, the text points with pride to the resources of the NUWC and to its achievements and awards. One of the document's several audiences is addressed in a message on the back cover, informing the reader how to inquire about employment opportunities at NUWC.

Two other capability statements, both issued by Air Force activities, follow a different style of presentation. They are picture books. *Wright Laboratory* (n.d.) presents 68 photographs on 28 pages to convey a graphic vision of the Laboratory's mission. That mission is to lead and focus the Air Force's aeronautical technology investment through in-house research and contractual partnerships with universities and industry. The mission statement is factored over a page of white space, accentuating four kinds of operations: discover, develop, transition, and solve. These operations become the category labels for groups of photographs of people interacting with machines. *Integrating Tomorrow's Technology Today*, issued by Ogden Air Logistics Center (n.d.), is also a tabloid presentation. Its vision, stated in two paragraphs, combines two claims: our creative and motivated people have pioneered in developing cost-effective repair facilities, and we have a broad range of capabilities to meet your needs. Groups of photographs illustrate the Center's seven categories of operation.

Insistently visual, these upscale tabloids raise the interesting question of how to convey a conceptual vision persuasively. Photographs appropriate virtually all of the page space; the presentation of the Ogden Air Logistics Center is even wrapped in a color photograph of the cosmos. The text is limited to captions, which are often inset into the photograph itself. These localized captions make supporting claims that are distributed throughout the entire capability statement. Continuity is superficial. Instead, the attempt here is literally to merge the vision with the evidence that supports it. The style does have its downside, which is information overload. The mind can absorb only so many pictures of technicians peering at computer screens or reaching into sections of sophisticated machinery. Like a spectator who has trooped too long through Disneyland, the reader suddenly crashes. Yet the design strategy is deliberate. Whatever else it achieves, it succeeds in representing the concept of human–machine interface as a kaleidoscopic universe.

NARRATIVE, PLOT, AND KNOWLEDGE PRODUCTION

Here is what we have discovered so far. The task of the capability statement is to yoke past evidence to a future claim. As in any argument, implementing this purpose entails making the evidence as relevant as possible to the claim. To this point we have seen several versions of a generic architecture that incorporates claim and evidence. A vision is declared and then extended to include periodic interpretations of information intended to verify it. The text is arranged to provide a sense of diffused support for the claim, and styles of animating the listed information vary with the intended audience. Efficient, bulletized lists provide adequate references for specialists, whereas graphic statements for lay audiences invite the reader repeatedly to interpret the information in terms of the vision.

The genre makes use of yet another knowledge-producing property which is narrative. As I have suggested, the power of narrative to convert a list into a coherent statement lies in the reader; coached by clues in the text, the reader does the converting. When the clues themselves are narrative fragments, the reader's inferences come easier. Yet narrative by itself does not guarantee persuasion. Narrative is simply chronology, whether backward or forward, and chronology is not by itself enough to get the reader to make the crucial inference that links past performances to future capability. To get the reader to make this leap, the narrative needs a plot.

There is a pivotal difference between a narrative with a plot and a narrative without one. By itself a narrative is merely a chronology of events; it simply says, "This happened, and then that happened." But a plot links those events in a causal way; it says, "This happened *because* that happened." In his *Aspects of the Novel* (1927), E. M. Forster has explained this difference with a mathematical elegance. Here is a narrative, he says: The king died, and then the queen died; and here is a plot: The king died and then the queen died of grief. Plot is the arrangement of events in a causal way; it is what makes a story. And story is what makes a narrative plausible. If narrative alone can link past events to a future vision, then narrative with a plot can weld them credibly together.

Two sustained capability statements, both issued by the Army, use plot to weld past events to future vision. They tell stories. Both documents are undated, but textual evidence points to 1992 or 1993. Both are printed in the same deluxe style of the other tier-1 capability statements, with many pictures, sparse text, and a lot of white space on every page. One of them is the *Army Technology Base Master Plan Overview* (Department of the Army, n.d-a). Although the task of this 24-page document is to describe the Army Technology Base Master Plan (ATBMP), its rhetorical purpose is to confirm the wisdom of this plan.

The vision is welded to its supporting evidence simply by the juxtaposition of two sentences. The text begins, "In order to win tomorrow, there must be a continued commitment to a robust, focused, and prolific technology base" (p. 2). This is the vision, and the next sentence contains a catalogue of evidence. It lists some of the technologies that prevailed in the Persian Gulf conflict: the Patriot missile system, Abrams tank, TOW missile, Apache helicopter, and smart munitions, all of which "resulted from the implementation of a balanced Army technology base strategy in the last 20 years." Cause-and-effect is empirically evident in these two contiguous sentences. The ensuing categories of supporting information pertain to the further development of the technologies that prevailed in the Persian Gulf conflict. These include tactical missile defense, armor/antiarmor systems, attack air vehicles, mines and countermines, the technologies that support command, control, and communication (C^3), and other technologies as well.

Late in the document, the text finally addresses the rationale of the master technology plan by presenting a schematic diagram. A vertical flow chart with boxes and arrows begins with the national security objectives and leads downward through six levels of organizational authority to the plan itself. By this time, the rhetorical thrust of the vision statement has long since won the battle for the reader's endorsement; the whole causal relationship between vision and evidence was self-evident at the beginning. In a kind of postscript, the text clears up one other matter. The final page reveals the social technology underlying this document. With a disarming reflexivity it lists 10 initiatives for supporting the Army's S&T plan; the 10th initiative is to publish this vision of it.

The other document I mentioned as incorporating the concept of plot is *50 Years of Accomplishments in Army Research and Development* (Department of the Army, n.d-b). This is an 80-page anthology of success stories. The 16 stories in the volume graduate the genre of capability statements to an art form. Each story dramatizes the use of a different technology that Army R&D has developed. The stories are a peculiar mix of fact and fiction. Events are arranged to convey a progression from actual R&D efforts that developed a given technology to some fictional instance of its use.

There is a twist in this progression of events, a narrative sleight-of-hand. Each story follows the conventional sequence of the generic capability statement; first comes the vision and then the supporting evidence. But in this case the vision statement is fictional; each story begins with a visionary extension of some actual Army R&D. Such a blatant tall tale would scarcely seem to convey the desired credibility. Yet it works. Each story begins with a dramatic fictional episode that presents an urgent predicament. Here is one such crisis:

> Tired and very frightened, she pulls her pink sweater even tighter around her shivering body. She huddles to a partially decayed tree, calling "Mommy. Daddy. Where are you?" The darkness is total now. An owl hoots in the tree overhead, and she begins to sob (p. 59).

So begins the episode entitled "Eyes Like a Cat," which continues to a narrative resolution. "I believe I've spotted the subject in Area 2," the voice on the walkie-talkie reports. Everyone in the ranger station leans forward to listen. "I'm almost there." Wearing his night vision goggles, the state trooper easily makes his way through the dense woods toward the 4-year-old. "Amanda? Amanda? If you can hear me, please answer." " ... Daddy? ... " "I've found her. I say again, I've found Amanda Baines" (p. 571).

So ends the episode, but the capability statement continues by explaining why it has a happy ending. During the 1950s, the Army began R&D on night vision devices, and continued efforts have developed two efficient, dependable technologies. One, called light intensification, can amplify ambient light 10,000 to 20,000 times; it is incorporated in night goggles, such as those that helped the trooper discover Amanda Baines. The other technology functions independently of light; called thermal imaging, it detects heat in the form of infrared radiation and produces recognizable images in the dark. After describing the successive generations of each technology since the 1950s, the text concludes by citing some outcomes of this R&D. Night-vision devices are now commonly used in law enforcement and drug traffic control, analysis of underground mine faults, and diagnosis of cancer.

"Eyes Like a Cat" follows the model of all the stories in the anthology. They begin with comparable—and fictional—crises that are resolved by the outcome of previous—and actual— Army R&D efforts. An accident on a major thoroughfare in the morning rush hour threatens to paralyze a city's streets until traffic engineers reprogram the traffic lights. The electronic brain that makes this possible is a legacy of the Army's Electronic Numerical Integrator and Calculator, ENIAC, developed 50 years ago to compute firing table calculations. The capability statement describes the evolution from this mammoth structure, the world's first electronic digital computer, to its present manifestations. Another story is apocalyptic. It begins with a world crisis over the acquisition of intelligence information from a satellite. Rogue nations are using the information to conduct lethal air strikes against United Nations peacekeepers, and the restoration of law and order is urgent. The crisis is resolved only when laser beams atomize the satellite. What follows this episode is an account of the Army's R&D in the evolution of laser weapons and electro-optic control systems, and their various uses in surgery.

The crises continue from one story to the next: The burned child, whose life is saved by immediate treatment at the Army burn center; a hungry soldier in the desert, who dines on a nourishing, portable meal; even the crisis of smallpox at Valley Forge, which is averted by vaccination. Like the crises in the other stories, these urgent episodes correlate evolving technologies, and make the capability statements coherent. They convey what rhetoricians call intention and what theater actors call a through line. Their dramatic perspective on the consequences of technology impart a distinctive interpretation of defense R&D and a specific kind of knowledge.

LET THE CUSTOMER BE WARY

The huge effort expended by these organizations to position themselves in the marketplace merits a passing question or two. The capability statements do produce knowledge; no doubt of it. Despite the trade-off in styles, they all interpret information by warping it to a vision of some sort. In so doing, they convey inferences about that information; they formulate new propositions, new instances of causality that were not originally inherent in the information itself. These changes add value to the information on which they draw. But a caveat may be in order.

Although *value added* commonly denotes a qualitative improvement, that assumption does not necessarily hold with respect to capability statements. As might be expected, the vision of a capability statement is a tunnel vision; it excludes some of the realities that beset defense R&D activities, such as underfunded laboratories, rusty equipment, or unstaffed projects. It is not the whole truth. Even the claims that *are* made can be chancy. A given juxtaposition of vision and information could be questionable or misleading or even fraudulent. Who would know?

In the absence of any consensus for judging capability statements in the defense R&D community, the worth of their *value added* is unknown. The transactional documents are different in this regard. They are subject to judgment by contract award, and whatever value they add to the information they shape is sooner or later accountable. But who monitors the claims in the capability statements? How valid are they? What referee function is at work? The persuasiveness of the new knowledge produced by capability statements, as with any argument, lies in the responses of the readers. These have yet to be measured.

Aside from the issue of honesty in advertising, one other caveat gives pause. Used by virtually all of the defense R&D activities, capability statements appear to be mandatory to the culture. They represent a large investment of resources. Given the sizable investment in trying to achieve credibility, therefore, it is surprising that there is no measure at all of its efficacy.

12

The Knowledge Cycle

Having explored the architecture of the major systemic documents in this discourse community, we can climb Olympus for a distant but more encompassing view of the way in which they produce knowledge. We are looking for a grand pattern here. Our context is the broad range of discourse that conveys the interactions of the defense S&T market. It includes all of the transactional documents that the past few chapters have been tracking, as well as the capability statements. As part of the acquisition process, the transactional documents are exchanged between customers and vendors in the ongoing design of R&D projects; they serve designated functions in the defense S&T market. The capability statements serve a different function in the market; they help position the participants for projects yet to come.

In the formation of the myriad R&D projects of this marketplace, these two kinds of documents continually overlap. There is no set order among them. Moreover, each kind of document produces a different kind of knowledge. To comprehend the interactivity of them all, a simple schematic view will help. In a highly abstracted sense, these two kinds of documents form a cyclical flow of knowledge from one project to another. One might conceive of a simple model based on the fact that every R&D project has its own life cycle. In this model the circulation of knowledge helps to sustain the concluding phase of one project cycle and the beginning phase of the next one. The notion of cyclical development helps to explain the large tidal flows of knowledge that occur in the defense S&T market. It provides us with an overview, which we turn to now.

What relates the transactional documents and the capability statements to one another is the rhetorical mode of problem and solution. In this discourse community it is the mode of choice; problem formulation is the accepted way of configuring goals and obstacles in order to gain new knowledge. Each formulation, of course, is parochial in that it represents a given point of view addressed to a given audience. Keeping in mind the model of cyclical flow, recall for a moment the documents that pertain to the process of procuring R&D:

1.–4. Needs (Initiated by the customer)
 Acquisition Forecast
 Sources Sought Announcement (SSA)
 Request for Proposal (RFP)
 Broad Agency (or Broad Area) Announcement (BAA)
5.–7. Offers (Initiated by the vendor)
 Sources Sought Response (SSR)
 White Paper
 Proposal
8. Contract (Agreed to by customer and vendor)

One recognizes these documents as redefinitions of a problem that either the customer or the vendor has originated. Through a dialogue of problem statements, the original idea is developed and modified by increments. Customers issue generalized forecasts of acquisition needs, followed by either BAAs or RFPs. In response to customers' solicitations, or on their own initiative, vendors issue white papers and then proposals. Through any of several document-exchange procedures leading to a contract, both parties reciprocate in refining a technical problem.

What we are seeing is a sequence of knowledge-producing events in the form of successive problem formulations. But the sequence extends beyond the procurement of an R&D project. To serve new purposes, both parties reframe the original problem in several different kinds of documents. Vendors deliver periodic reports to their customers, and, after the conclusion of a project, they may also publish articles or prepare patent applications; they address different audiences for different purposes. To support new marketing initiatives, moreover, vendors also generate the kinds of capability statements we have been reviewing; their purpose is to recast the original problem into explanations of their new capabilities. And to accommodate their own marketing needs, customers do the same. In the dynamics of the market, capability statements issued by both customer and vendor complete the cycle of knowledge production related to any research project.

MAPPING THE PROBLEM CONSTRUCTS

The problem construct is the knowledge-producing engine in this discourse community. It is what organizes all the generic documents in both phases of the knowledge cycle. As we have seen, this construct has a characteristic architecture. But there is yet more of it to explore. The notion that texts contain large, discrete structures of information—structures that contain groups of paragraphs—has been well developed in rhetorical studies; it has been carefully analyzed by British specialists in applied linguistics. Notable among these scholars are Michael Hoey (1983) and Michael P. Jordan (1984). Hoey has delineated a syntax of large, constituent structures of English-language texts, and Jordan has designed a strategy of pattern recognition based on such structures. Jordan proposes that comprehending the line of reasoning in any given text lies in recognizing the pattern by which these information structures are combined. Specifically, the strategy of pattern recognition is based on the sequence of component structures and on the kinds of mechanisms that signal their presence.

According to both Hoey and Jordan, a basic metastructure in English texts is some combination of four discrete information structures, designated as:

1. Situation
2. Problem
3. Solution
4. Evaluation

The structures can be illustrated in an English text on virtually any subject. We might compose a text, for example, on the subject of federal flood insurance. It would be a metastructure. A beginning set of paragraphs might describe a *situation* in terms of hurricanes that come ashore on the east coast of the United States. A second information structure, or set of paragraphs, might pose the *problem* of protecting the equity of homeowners in this situation. A third set of paragraphs might describe the federal flood insurance program as one *solution* to the problem. And a fourth set of paragraphs might present an *evaluation* of the costs and benefits of that program.

This set of information structures can be arranged in different sequences to suit different purposes. What might we want to emphasize? Perhaps the changes needed in the present flood insurance program? Or the cyclical nature of hurricane frequency? Or the different kinds of solutions to the problems of home equity in this situation? They are all possibilities, and they all depend on how we arrange the information structures. The architectural features that distinguish one text from another are the number of information structures it contains and the sequence in which they appear. Pattern recognition consists of tracking these two variables.

Hoey and Jordan are dealing with patterns within a single text, but their strategy of pattern recognition can be extended to intertextual dimensions. The extension of this strategy happens to explain the relationships among the texts that circulate in the defense S&T market. To discover a prevailing pattern of knowledge flow in the market, we might look to the documents that correlate it. We can extend this strategy of pattern recognition to the texts of the whole corpus of documents that we have been viewing.

We already know that these texts are related in a general way, and we know that the transactional documents are specifically problem oriented; they reveal an evident pattern of information structures. So incorporating the texts of capability statements into our intertextual perspective should provide us with a prevailing pattern of knowledge flow. It should give us a way of describing the successive problem formulations in the life cycle of an R&D project and, at the same time, a way of visualizing how the problem formulations of different projects might intersect.

We are about to see this pattern worked out graphically. Although the transactional documents contain different information structures, they all reveal one remarkable similarity. They all contain some combination of problem and solution. These are the foundational structures of a pattern. The other two structures (situation and evaluation) are ancillary. Once this fact becomes evident, it is relatively easy to discover how knowledge flows through this discourse community. The successive documents of an R&D project do not present different problems. Instead, they present different versions of the *same* problem. The various problem formulations are actually reformulations. They are different combinations of the same information structures arranged to accommodate either a different purpose or a different audience. Each successive reformulation generates an increment of knowledge that develops or modifies an original technological idea.

This process of rhetorical reformulation is not limited to the transactional documents. It extends beyond them, to include the capability statements as well. To be sure, the transactional documents have the more highly developed texts, and their sequential use generates more occasions to reinterpret a given problem. Yet the capability statements are also part of the life cycle of the R&D project, and they are also interpretative. The whole reach of the process includes them.

We can map all of the generic documents in a way that shows their structural relationships to each other. In the form of a matrix, Table 12.1 illustrates typical combinations of the four information structures that occur in the formation and resolution of an R&D problem. The constituent elements, or information structures, are listed across the top of the matrix. Down the left-hand column are the generic documents that move the R&D project forward to its conclusion. And the Xs located throughout the matrix mark the occurrences of each information structure in each of the generic documents.

The table reveals three systemic features of the discourse in this community. One is the proportion of total documents issued by the customer and by the vendor. (Those documents appear in the two left-hand columns). Another feature (reading across the matrix) is the number of information structures (marked by an X) in each document. The third feature (reading downward) is the distribution of those structures in successive texts. This distribution is a graphic illustration of a whole contextual pattern of discourse.

Table 12.1 displays two prominent features of the intertext. First, most of the texts use only two of the four available information structures, problem and solution. Either the information for those other, unused structures has

TABLE 12.1

Information Structures in Defense S&T Documents

Author of Document		*Problem Elements:*			
Customer	*Vendor*	*Situation*	*Problem*	*Solution*	*Evaluation*
Acquisition Forecast		X	(X)		
Broad Area Announcement		X	X		
Sources-Sought Request		X	X		
RFP		X	X		
	White Paper		X	X	
	Sources-Sought Response		X	X	
	Proposal		X	X	
Contract	Contract	(X)	X	X	
	Final Report		(X)	X	X
	Patent Claim		X	X	X
	Journal Article		X	X	X
	Capability Statement			X	X

Note: X denotes typical usage. (X) denotes occasional or unstressed usage.

not been generated at the time of composition, or the reader is presumed to be familiar enough with such information for the writer to omit it. Second, and more important, is the unmistakable drift of those Xs toward the right side of the matrix. That drift reveals an incremental progression of the paired structures. As the life of the project continues (downward in the matrix), successive documents emphasize different pairs arranged in the same sequence: first situation and problem, then problem and solution, and finally solution and evaluation. In other words, successive iterations recontextualize the original problem, moving it through stages; each iteration changes the form of the problem and its relative significance throughout and beyond the life cycle of the project.

Table 12.1 confirms the drift of the problem-oriented discourse that I have been describing. In the procurement of an R&D project, a perceived need is gradually formulated into a problem to be solved. Confirmed through a series of transactional documents exchanged between customer and vendor, this formative process leads to the design of a project, a fundable enterprise. What is actually funded is a problem that is deemed to be solvable. After it has been solved, the problem is then reformulated for other purposes or other audiences.

CYCLICAL FLOWS OF KNOWLEDGE

This pattern of knowledge production can be likened to the shape of an hourglass. To repeat the pattern in these analogous terms, knowledge flows like sand through an hourglass, from the broader discourse community to the participants in an R&D project. It is funneled into a specific statement of work as specified in a contract, and then dispersed to the broader community. Flow is the functional metaphor borrowed from economics. In Machlup's theory (1980), the commodity of knowledge exists in both stocks and flows, but only the flows are productive. The dissemination of knowledge from one stock to another is what produces knowledge. In Fig. 12.1, an hourglass delineates the channeling of this flow from one stock of knowledge to another.

In the first phase of knowledge production, prior to contract, either customer or vendor invokes some set of technical ideas whose possible relationships are as yet indistinct. The ideas are formulated into a problem, and successive iterations develop it to the point of designing a research project. In this first phase of its production, knowledge is progressively narrowed and focused; it becomes stabilized in the research design.

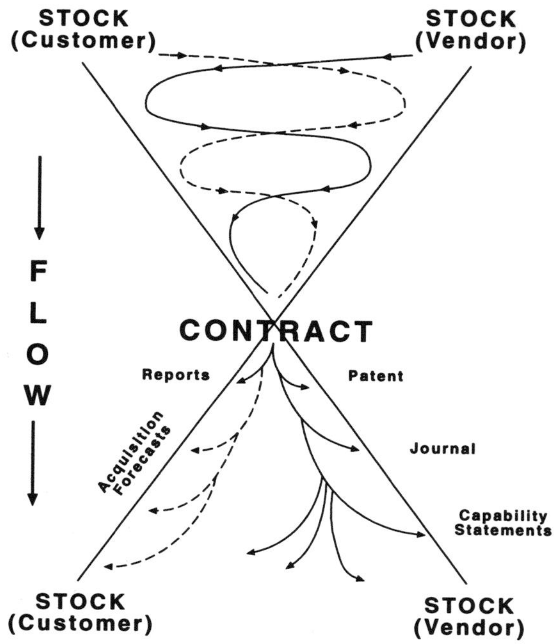

FIG. 12.1. Stabilization and destabilization of knowledge.

After the project concludes, the knowledge that the R&D yields is systematically disseminated. In this latter phase, like sand in the bottom chamber of the hourglass, knowledge quickly becomes destabilized. The vendor writes reports to the customer, then later to a larger research community, in the form of articles for trade publications or learned journals. Or, if the findings of a project have equity in the market, the vendor might file for patent protection instead of publishing them.

Published articles and patent applications have different communicative purposes, and their texts are differently configured. Yet they are alike in that each is highly dependent on its social context.

Formal requirements for patent applications stipulate the inclusion of certain kinds of information, as well as an explicit rationale for the claim being made. Convention levies the same constraints on the author of a journal article. In each case, the claim is a painstaking interpretation of the information.

Although they are responsive to different kinds of authority, published articles and patent applications generate knowledge in highly formulated ways. The patterns of dissemination in each case are well established. Through different channels, both publication and the patent process destabilize the knowledge that was concentrated in the R&D project.

Along with these other means of dissemination, the vendor that conducted the R&D project addresses prospective customers in capability statements based partly on the new knowledge. And the sponsor of the project, now turned vendor, does the same. To help position itself in the market, the sponsor puts its own spin on the knowledge output of the project and then disseminates it. That knowledge, moreover, will probably be used as background explanation in future acquisition forecasts.

Problem formulation shapes the flows of knowledge in all phases of an R&D project. It focuses the flow prior to the project, stabilizes it during the project, and shapes its dissemination afterward. The hourglass figure is a static display of an ongoing process in the defense S&T market. The successive stabilizing and destabilizing is a tidal flow. We can sense the volume of the flow by visualizing it on the scale of an ocean. Hundreds of thousands of current R&D projects are underway in different stages at any one time, and the hourglass depicting one project expands by that many orders of magnitude. In this giant hourglass, each knowledge cycle has a fair chance of affecting other cycles in some way at some time. The destabilizing of knowledge in the latter phase of one research project may accelerate or modify the earlier, stabilizing phase of some other project. Such is the expansive property of problem formulation in this ocean of evolving knowledge.

In addition to the volume of flow, one other inference can be drawn from the hourglass figure. It concerns a lingering question over the relationship between information and knowledge in the flow process. Machlup acknowledged the blurring of the two in common usage. But he chose not to press the technicality, since it did not interfere with his primary concern for the value of knowledge as a commodity. Yet the distinction between information and knowledge in the flow process helps to explain more precisely how knowledge happens. The flow consists of both information and knowledge, and the flow process consists of successive conversions of one to the other. Since these conversions are particularly relevant to the enterprise of technology transfer, which we are about to address, some brief clarification is in order.

Information is generic. Like mass and energy, it exists in many forms. In its most atomistic form it consists of data, separate signals or emissions with no particular relationship to one another. In its most organized form it is characterized as knowledge. Once this distinction is made, the dynamics of the flow process becomes clearer. With this in mind, we can invoke the hourglass figure again. The successive stablizing and destabilizing that it displays pertains *alternately* to information and knowledge.

The alternation is cyclical. As customer and vendor move toward an R&D project, information from their several perspectives becomes more organized. When it is stabilized by the mutual understanding that a contract confirms, the information becomes knowledge, and the project itself refines it. Then, through its dissemination after the project concludes, that knowledge acquires a different status. Destabilized, it becomes information, albeit new information. No longer organized exclusively for the project that engendered it, this new information is accessible for more and different interpretations. And these subsequently yield new knowledge. Through the alternative phasing of information and knowledge thereby, knowledge production is a regenerative process.

Coming to Terms IV

We have been exploring some major elements of a vast and complex knowledge factory. They include a canon of critical technologies, a system of burdensome regulations, a paper trail of documents, and a particular mindset that formulates solvable problems. We have been viewing these elements from a variety of perspectives: knowledge theory, political economy, problem theory, rhetoric, and marketing. These perspectives merge into an evolving perception of a system of knowledge production that is now undergoing profound change. That change is the phenomenon of defense conversion, which we are about to explore in the rest of this study. We need to come to terms with it.

The elements of economics, politics, and rhetoric that we have so far been analyzing all intersect in this new phenomenon, which has problematic consequences for the production of knowledge. They are problematic because the national mandate for defense conversion entails doing business in a radically different way. There is no precedent, no guide for going about it. The situation has been regarded by many as a crisis. In military parlance, it represents a new challenge.

The need for a massive correction in defense spending became evident shortly before the end of the Cold War. The flow of resources to the defense technology base had rapidly accelerated. Prompted by the defense buildup in the 1980s, and especially by the Strategic Defense Initiative, the flow became so imbalanced as to warrant radical change. In summarizing the urgency of the need for change, Peter Drucker (1989) spoke for many: "The first economic priority must thus be to reduce and bring under control the defense spending and the diversion of high-grade human resources into economically unproductive—indeed, economically counterproductive—defense work" (p. 50). Moreover, the alarm was not only over the drain on the national economy, but as well over the erosion of resources for developing technology. As documented by the congressional OTA (1989), the Cold War method of investing in defense was causing defense technology to lose its strategic effectiveness.

A broadscale defense conversion is the proposed antidote to this crippling condition. The vision of this new phenomenon is twofold: to reduce sharply the money available for defense resources, and to use present DoD resources to help build a more competitive national economic infrastructure. For the development of defense technology, the vision entails more than merely transferring fiscal resources from one category of the economy to another. It entails a new infrastructure, a merger of national resources into a single, comprehensive technology base.

Merger is the key word here. Elsewhere, I have spoken of the two technology bases that support national defense that are commonly referred to as being discrete. As the congressional OTA (1988) described them, one is national; large and comprehensive, it is a function of the economy as a whole. The other is smaller and more acutely focused; it is within the domain of the Science and Technology Program of the DoD. For nearly 50 years, the knowledge flowing between these two technology bases has been directed largely by defense needs. Now comes defense conversion, the envisioned enterprise for integrating them.

But it is a distributed vision, marked by contradiction and revision. In 1993, the Clinton administration embarked on a set of initiatives in technology policy, and two years later a Congress of opposite persuasion abruptly defunded them. Although they have been virtually bankrupted, they still have identities as programs. One of the initiatives, the Technology Reinvestment Project (TRP), was designed for the development of dual use technologies. Headed by the ARPA, in conjunction with four other federal agencies, the program is meant to enable DoD and commercial firms to engage in partnerships for both R&D and manufacture. ARPA (1995) summarizes the rationale and the scope of the program. Another initiative is the Advanced Technology Program (ATP) of the National Institute of Standards and Technology. It entails federal funding jointly with commercial firms to insure those firms partly against the risks of innovation. The partisan politics with respect to these programs is a story unto itself, one which has little to do with the merits of defense conversion. Nonetheless, by fits and starts, the concept has taken form through various administrative and legislative mechanisms for nudging the two technology bases closer to each other.

One forthright measure for merging resources, as well as cost-cutting, is DoD's elimination of military specifications for its technologies. These are the engineering and technical requirements that contractors have had to follow in manufacturing weapons and equipment. The distinctiveness of MILSPECS lies in what they stipulate. In addition to describing requirements for the performance of a device or a system, they specify the procedures that the supplier must follow in manufacturing and testing it. MILSPECS are the result of overdesigning a device in order to achieve its total reliability.

This costly injunction governed the procurement of defense technologies until 1994, when Secretary of Defense Perry announced its demise as part of the defense conversion movement. At that time there were 31,000 MILSPECS stipulating how devices are to be built and tested. Thereafter, program managers in defense activities would be limited to telling contractors what they want to buy, but not how it should be built. *Defense News* broke the story of Perry's historic announcement. In its pages, Stephen LeSueur (1994) quoted an analogy about organizational culture made by Greg Saunders, a deputy in the DoD office charged with implementing the change: "We have a huge culture of a guy sitting in the middle of New York City beating a drum to keep the lions away. It seems to be working, so he is not going to stop beating the drum" (p. 23).

The conversion to performance specifications has already affected the cultures of both customers and vendors. DoD has undertaken extensive training for program managers in how to rethink what they want to buy. Defense contractors are revising their organizational structures in order to merge their commercial divisions and their defense divisions, which have been diligently separated for years. Consulting firms are advising commercial firms about their market opportunities, based on new kinds of business arrangements with the federal government. The seminar business is flourishing.

As for the defunded initiatives that I mentioned a moment ago, the TRP involved federal agencies and commercial firms in various types of partnering arrangements that include both R&D and manufacture. Despite TRP's limited budget, the partnering arrangements on which it is based were authorized by earlier legislation, and they are now widely used. One of these is a Cooperative Research and Development agreement, the CRADA. It is a compact between a federal agency and a commercial firm, which has some properties of a grant and some properties of a contract. Another arrangement for partnering is the Cooperative Agreement, which differs from the CRADA in that it allows the government to invest money in the partnership.

The other defunded program, the ATP, offers a different kind of agreement between the government and commercial firms. This venture employs joint funding as a kind of risk insurance. In one form or another, notwithstanding partisan politics, all of these arrangements are likely to surface again, because they benefit all parties to the agreement. The Defense S&T Program needs access to the state-of-the-art knowledge products in the commercial sector, and commercial firms need the subsidy for high-risk ventures.

TECHNOLOGY TRANSFER

Other legislative programs to come will probably help to implement the basic concept that underlies defense conversion. The concept is technology transfer. Its rationale is to make the economy more competitive, and one of its intended benefits is that federal laboratories will help pay their own way. Its strategy is to create new knowledge flows from the DoD technology base back into the national technology base. Technology transfer is the subject of the final chapter of this study. It is the most durable element of defense conversion, and it involves by far the most widespread participation among all sectors of the economy. So our coming to terms with it is timely.

Technology Transfer is another one of those abundant abstractions. In casual usage, which is frequent, the term connotes the transmittal of information or knowledge from one party to another. Some people shrug it off as simply what happens when one person communicates with another about something technical. Others use it in a historical sense to describe the general flow of technical knowledge from the European colonial powers to their colonies. In this sense, the term has been used to explain how textile manufacture migrated from England to New England. In the same usage, it denotes how the chemical industry in Germany subsidized the development of the chemical industry in the United States.

As an enterprise involving the federal government with other sectors of the U.S. economy, however, technology transfer has a much sharper focus. People who work every day at the task of transferring technology have in mind an explicit operational definition. To them, *technology transfer is the dissemination of knowledge developed within one organization for one purpose to another organization for another purpose*.

This operational definition focuses on the knowledge-producing process, specifically on the change of purpose for whatever knowledge is transferred. But it leaves open the kind of organizations that might be engaged in the transfer. The organizations might be two different divisions of the same corporate entity; they might be the sales division and the engineering division of General Motors, for example. And the transfer of technologies from one to the other might occur in either direction. Or the organizations might be two different corporate entities in different sectors of the economy; technologies might be transferred between a software company and a steel mill or a brewery. Within the defense conversion movement, however, the people engaged in technology transfer are primarily concerned with knowledge that has originated in federal agencies and that can flow in all directions: to other agencies, to state and local governments, and to commercial companies.

Disseminating old knowledge from its niche in one organization to a new niche in another is not simply a matter of dispatching it. The operation is far more intricate. Consider, if you will, merely the mechanical procedures involved, without any concern, for the moment, over why the transfer is being made. In the first place, what will become knowledge is not yet knowledge at the beginning of the transfer operation. What will become knowledge exists as inert information, trillions of bytes and billions of pages of it. Someone has to know what it is and where it is stored. Then someone has to describe what it is and how it can be accessed, and then someone has to publish that new information about the inventory. Thereafter, someone has to interpret from the inventory of information what parts of it can accommodate a given need or purpose, thereby converting it to knowledge. Finally, someone has to access the relevant information and download it into another file. From there it is converted to a different application, becoming new knowledge.

Theoretically, two parties are involved in this transaction, the information custodian and knowledge user. Actually, a great many specialists of one sort or another are involved along the way in merely the mechanics of the transfer. The communication between the information custodian and the knowledge user, however, is another matter. The custodian and the user represent organizations that have different goals and obligations. Their needs and expectations are not necessarily the same, and neither party is likely to know how to translate those needs and expectations to one another without a good deal of interaction. The situation is new to both of them. The custodian asks, "What do you want? I have a lot of it here." And the prospective user answers, "How do I know what I want until I see it?" The dilemma they share is a classic illustration of the need for marketing.

One symptom of this dilemma surfaced in a conversation at a meeting of defense laboratory directors, organized under the Joint Logistics Command. The scene was a mezzanine in a hotel near the Pentagon, during a break between meeting sessions. The participants had just been hearing about the labor-intensive nature of a technology transfer problem in Kansas City; they were commenting on it with the familiar candor of professional colleagues. Noting the costs involved, one participant, a laboratory director, asked, "Why not just get a report out of the files and send it to Kansas City?" Another participant, a technology transfer consultant to a different laboratory, responded, "If reports don't penetrate your own organization, how do you expect them to do any better out of town?" A third participant, an admiral who had been listening, said to the consultant, "You know, you might have something there."

The dilemma over how to formulate technology transfer is implicit in this conversation. The laboratory director missed the whole point of technology transfer, the consultant already knew the point, and the admiral was just becoming aware of it. Although the supply of potentially useful technologies that can be transferred to the market is inexhaustible, neither party to the transfer is a marketer. The final chapter of the book engages this dilemma of a default in marketing. We can already anticipate what to expect in a transfer situation where marketing is not part of the interaction.

CORPORATE CULTURE

The conversation among the laboratory director, the consultant, and the admiral brings up one other critical factor in the problematic nature of technology transfer. Each one of these participants who saw the problem so differently represented a different corporate culture. This is one more concept that we need to come to terms with in exploring the dynamics of technology transfer.

The information custodian and the prospective knowledge user work in different cultures. Their organizations have different values and different rules of procedure; they have different business priorities. The organization of the information custodian has a new mandate to transfer technology, but it also has an older, and very different, mission which is its reason for being. Its mission is to develop or acquire knowledge related to some designated field of operation. A mission, like a charter or a franchise, specifies the purpose of an organization and, therefore, its priorities and principal obligations. Now suppose that there is only limited funding to support all the obligations of the custodian's organization. How can this situation lead to a solvable problem? The dilemma helps us to recognize the strictures of corporate culture.

It is commonplace that every organization has a culture, a set of values and a characteristic way of doing things. It is also commonplace that this culture is conveyed to the organization's members or employees through informal rules of behavior that convey how they will be rewarded or punished for their efforts. It is conveyed by familiar but powerful rituals that discipline people: by the assignment of office space and by the ways in which meetings are structured. How are the seats arranged? Who sits where at the table? Who has the privilege of stating a contrary opinion, and who does not?

The rules of these protocols are nowhere written down, but they are imperious. Corporate culture is conveyed by dress codes, by who lunches with whom, by the various responses to an employee's initiatives; it is conveyed by what people talk about at the coffee breaks, and by the kinds of defensive postures people put into their memos to one another. These behaviors are all distillations of a value system by which that organization lives. The values are derived from its purpose or mission, interpreted by its managers, and passed on down the line.

Corporate culture is a powerful inertial force; it affects the rate at which an organization can change, and this fact bears significantly on the ways and means of federal technology transfer. Although each federal agency has its unique culture, there are some broad patterns in how managers interpret their missions. It makes a cultural difference, for example, whether a federal R&D organization is GOGO, government owned and government operated, or GOCO, government owned but operated by a private contractor. The directors of each type of organization are likely to interpret differently the best ways of meeting priorities. All of DoD's laboratories are GOGOs, for example, as distinct from R&D laboratories in DOE, most of which are GOCOs. Both agencies are subject to the same mandate for technology transfer, yet each has a different style of implementing the mandate. Managers must decide what is politically feasible and which battles they want to win. For one thing, the mandate to transfer technology is more intense at DOE, where the alternative is to go out of business. For another, the GOCO organizations of DOE are more marketing oriented about the transfer transactions that I have been describing.

As in any new venture, there are organizational risks in technology transfer, and rewards as well. How the risks and rewards are interpreted and how the mandate is carried out, profoundly affect the production of new knowledge. Within the defense R&D activities, such actions have seriously affected the process of defense conversion.

13

Reaching Out

In the preface to one of his novels, Henry James describes melodrama as an abrupt discontinuity in the coherence of a narrative; it is characterized by an event that previous events have not prepared the reader to understand or accept. In this sense, the military engagement in technology transfer is melodramatic. During most of the Cold War, the defense R&D activities served a culture that avidly acquires knowledge and then carefully locks it up. Then, with little or no preparation, they were ordered to transfer most of this knowledge to civilian organizations.

In a series of public laws enacted since 1980, Congress has directed some 600 federal laboratories and R&D centers to make their technologies available to one another and to other corporate entities. Those entities now include state and local governments, universities, and private corporations. The authorized means of transfer are virtually unrestricted; they range from individual consultations to corporate liaisons featuring liberal patent license agreements. Although directed to all of the federal laboratories, the mandate causes particular trauma for most of the military R&D activities. Having been knowledge consumers for 50 years or more, they must now also become knowledge suppliers. From their perspective, technology transfer would appear to be irrelevant to their missions.

Moreover, the task itself is daunting. If it were to consist merely of opening files and making patents accessible, not very much technology would be transferred. The success of the enterprise depends on information suppliers reaching out to potential knowledge users and interacting with them according to their needs. And since most of those needs will be imprecise at the outset, information suppliers will have to help formulate them.

The multilevel task requires a new social technology, one that combines several discrete skills. The tool-level skills of information storage, retrieval, and dissemination are necessary components, but by themselves they are not enough. To be effective, the new technology entails an ability to perceive a problem from the client's frame of reference. It also entails knowing how to formulate that problem in ways that convert information to knowledge. At present, very few defense R&D activities have the resources to develop this kind of informed effort. And with the restricted funding imposed by the defense drawdown, they are not likely to.

This chapter addresses these issues. Technology transfer has significant economic consequence for defense conversion, and effective performance of the mandate is the goal declared by DoD and the military departments. Yet the operational response to the mandate has been uneven. Here are the elements of what Newell and Simon (1972) call a well-defined problem. The obstacles to achieving the goal appear to lie partly in the task itself, and partly in the corporate culture that shuns it. A solution depends on the development of a viable social technology that can cope with these obstacles. To explore this problem, the chapter begins with an overview of the federal technology transfer movement. Then it examines the obstacles to achieving the goal and, finally, an emerging solution.

THE INFRASTRUCTURE FOR TECHNOLOGY TRANSFER

The federal technology transfer movement began in the Department of Defense in 1970, in response to congressional encouragement. Hearings before the House Subcommittee on Science, Research, and Development two years earlier had generated enthusiasm for broadening the use of federal laboratories. The subcommittee's view was favorable. In a flagging U.S. economy, the prospect of generating income from the sale of federal patents was attractive. Technology transfer was seen to be a feasible source of revenue in a domestic market with paying customers. The basic idea was to recycle technologies developed in DoD, NASA, and the Atomic Energy Commission (AEC). DoD's response was swift. A plan called the Domestic Action Program was drawn up and immediately implemented by the Naval Weapons Center (NWC) at China Lake, California. NWC needed extra funding to support long-term research, and the new program provided an opportunity for NWC to sell its capabilities to other federal activities. From its outset, therefore, technology transfer was market-driven.

Not surprisingly, marketing played a crucial role. NWC retained Perrin Associates, a consulting firm interested in technology transfer, to represent the idea to both suppliers and potential customers. The suppliers were the scientists and engineers who were developing technology at NWC, and the potential customers were any other federal activities that might be able to use recycled technology. Perrin also canvassed other DoD activities to study the possibility of forming a consortium of suppliers. These marketing efforts were endorsed by DDR&E, and in 1971 the DoD Laboratory Consortium was formed. Representatives of 15 DoD laboratories met at China Lake and elected to send a liaison officer to NSF to begin to link suppliers with customers. Both the DoD and the Navy issued policy statements supporting technology transfer and cooperative development. The rapid succession of all these events was remarkable. Even more remarkable was the ready support of the venture by DoD, DDR&E, the Navy, and NSF. The cooperative interaction of the market and the state in this instance was a rare occurrence, and marketing catalyzed it.

Other blessings quickly ensued. In March 1972, President Nixon sent a message to Congress endorsing the idea of domestic technology transfer. The GAO reported favorably on ways of extending defense technologies for civilian use. DDR&E favored extending the DoD Consortium to all federal laboratories, and in 1974 the Consortium members issued the invitation. Shortly thereafter NTIS agreed to provide market development support. But these blessings did not come automatically. A few heroes at China Lake, at NSF, and in DDR&E worked strenuously to make them happen. And more strenuous work lay ahead. It would be another six years before Congress required the Federal laboratories to participate in disseminating technical information, and yet another six years before it chartered the Federal Laboratory Consortium (FLC) to function nationwide.

These events are chronicled by Leroy Doig (1994) in the *FLC History*. Its major theme is the strenuous effort that was involved in securing an acceptable legal status for technology transfer. The enabling legislation came in installments, which are listed in *Technology Innovation* (1994), an annotation of the public laws in Chapter 63, United States Code. The first installment, the Stevenson-Wydler Act of 1980, required federal laboratories to participate in technical cooperation. For coordinating such cooperation, it also established Offices of Research and Technology Application (ORTAs) within the major laboratories. But the law did not fix any executive responsibility for seeing that technology would be transferred, so efforts to extend it began almost immediately. The Federal Technology Transfer Act of 1986 somewhat repaired this oversight by assigning to all federal laboratory scientists and engineers the responsibility for transferring technology. It also chartered the FLC, giving it implicit executive responsibilities. Finally, it provided a minimal funding mechanism by directing that a small fraction of each laboratory's annual R&D budget be used to support technology transfer. By 1993, 11 more public laws and two executive orders extended the enterprise piecemeal.

THE RESOURCES FOR TECHNOLOGY TRANSFER

The 1986 legislation chartered the FLC as a tier-3 federal agency within the National Institute of Standards and Technology (NIST) of the Department of Commerce (DOC). Thus empowered, FLC developed the organizational infrastructure that now guides the patterns of federal technology transfer. It became a confederation of six regional centers that have their own ties to state and local governments, commercial firms, and industrial associations.

FLC has also developed close organizational relationships with both the DOE and NASA. Its comprehensive electronic networking capability uses the resources of DOE's Lawrence Livermore National Laboratory; with this capability, it maintains a national online directory of subject specialists. DOE's Los Alamos National Laboratory is FLC's administrator and procurement agent. When Congress authorized the National Technology Transfer Center (NTTC) through NASA's budget, FLC helped to establish it. NASA and FLC jointly sponsor training programs and share networking resources. FLC's network also includes nonfederal organizations, representing industrial, governmental, regional, and university interests in technology transfer.

The infrastructure for networking federal technology transfer is extensive and complicated, and the continuous flows of information from one node to another flood the network. But there is a remarkable simplicity about the way the flows are channeled; they move between the information suppliers and prospective knowledge users. Building the channels and maintaining them, FLC is a full-scale broker. It provides not only the channels, but also access to the subject specialists, who can help convert the stored information into knowledge. Kathleen C. Hayes (1992) has catalogued these resources in Tapping Federal Technology. And Mary Stevanus (1994) has composed an online manual for linking the Federal laboratories with other enterprises.

The information sources are copious. The annual budgets of the federal laboratories total some $25 billion, or about a third of the total federal R&D investment. An estimated 100,000 scientists and engineers work in the laboratories, representing about one of every six scientists and engineers in the country. Their accumulated technical experience during the past 50 years totals many millions of staff hours, and their output represents a huge technical stockpile of patents available for licensing. Included in the stockpile is output from about 90 R&D activities within DoD, and from the 10 FFRDCs dedicated to defense R&D. In addition, DoD contracts with about two dozen Information Analysis Centers (IACs), which collate technical data pertaining to the canon of critical technologies.

The proportion of these defense technologies that have commercial value is unknown; informed estimates range widely, from 30% to 80%. The low estimate is based on current state-of-the-art technologies, and the high estimate is based on the total inventory accumulated over the years. Some of the stored technologies, will never be used commercially because they are already out of date, such as patents that pertain to analog computing. But many others are still evolving. What *is* known is the bewildering range of technologies, from the familiar to the esoteric, that have already been recycled. They include magnets and mirrors and paints, as well as high-level computer languages, molecular modeling, simulation software, and processes involving optical lithography and digital speech coding. Even highly specific military technologies sometimes have a specific commercial payoff. Guns and explosives would seem to be low on the list. Yet the barrels of some large-bore artillery rifles have ceramic liners that significantly decrease friction in the cylinders of automobile engines. And the Army's Picatinny Arsenal has miniaturized explosives to the point where a medical team can use the technology to blow up a gall stone.

Food technologies are more obvious candidates for new markets. As Mike Toner (1995) reports, technologies developed at the Army's Natick Research, Development, and Engineering (RD&E) Center have created new demands. The blending of proteins, fruits, and complex carbohydrates into portable packages has been tested and found far superior to candy bars for sustaining high energy levels. Developed for soldiers, such foods are also needed by forest firefighters, air traffic controllers, and hospital emergency room personnel. The challenge is to convert these needs into commercially feasible demands.

THE DIFFERENCE BETWEEN PROMISE AND PERFORMANCE

One reason for the imprecise estimates of the applicability of the stored technologies is the lack of information about the needs of potential users.

Overcoming this lack, as commercial vendors have long known, entails market research, which is the first step in marketing. Although proven to be cost-effective in commercial markets, preliminary market research requires front money and organizational time. But these resources have not been made available in most federal laboratories. Even more basic than the needed resources for market research is the need to understand that such effort is necessary.

There is a quiet irony in this situation. The abundant capability statements issued by the defense R&D activities signal that all is well with federal technology transfer. Most of the activities have printed documents that declare their readiness to participate in technology transfer. The documents range in visual quality from a single printed page folded for mailing to elegant books with four-color printing on glossy paper. They all have the same substance. The laboratory's mission, its facilities and resources, and a list of its impressive achievements form the core of these declarations of expertise. In typical repackaging, the shorter capability statements have been slipped into display folders that headline the idea of technology transfer, along with the laboratory's logo. In the longer statements, the front matter has been revised so as to feature the same idea. Most of the documents contain some reference to patents that are available for licensing, with information about the persons to contact. Clearly, the laboratory print shops are ready for technology transfer. But in the exhibition booths at trade shows and on the display tables at association meetings, their graphic products are typically stacked alongside a sign inviting browsers to help themselves. Sometimes there are no signs.

The amended capability statements testify to the formal commitment of DoD and the military departments. There are records of conspicuous success; careful count is kept of the numbers of CRADAs that engage R&D activities with civilian organizations, and the results are monitored and circulated in-house. Many instances of successful collaboration between laboratories and their clients have been documented. Winners in Technology Transfer (FLC, 1994) lists some of the more definitive results, and FLC and NASA regularly feature new achievements in their newsletters.

But the amended capability statements are also symptoms of a continuing predicament; they testify to a formal commitment by upper management that is not wholly honored at the operational level in the laboratories. In 1990, four years after passage of the pivotal Federal Technology Transfer Act, members of the Technology Transfer Society confirmed this situation. In a poll, reported by Joseph W. Lee (1993), there was conclusive agreement over the impediments to the federal technology transfer program. The indictment was categorical:

- Government has not been effective in informing industry of its technologies that have potential commercial value;
- Government lacks technology management expertise for technology transfer;
- Government bureaucracy is too cumbersome and difficult to deal with; and
- Cultural differences between government and industry have negatively influenced their working relationships.

This was the prevailing opinion in a nonprofit professional association. Four years later, the same indictment was still forthcoming, this time from a different sector. Randy Barrett (1994) reported the reluctance of commercial brokers to do business with the federal laboratories. Based on opinion sampling among commercial technology transfer brokers, the report states that "federal labs still suffer from a widespread reputation for passivity and red tape" (p. 36). As reported, the consensus of the CEOs of these brokerage firms was that the laboratories take a reticent approach to technology transfer. The causes of the reticence, they believed, are both political and cultural. One CEO offered the opinion that the scientists in the federal laboratories are simply not accustomed to dealing openly with the public.

The evidence supports a different conclusion. The defect appears to lie with the system, not with the scientists. Granted, more than 10 years after passage of FTTA, only about half of the defense R&D activities are participating in the FLC network, despite the directives to do so. But the obstacles to satisfactory performance appear to be embedded within the layers of management, and the symptoms are evident.

The FTTA requires the establishment of ORTAs in each laboratory with more than 200 employees; it requires that this function be funded as part of the activity's R&D budget. But some ORTAs are regularly short-funded. Administrative support for the ORTAs, moreover, is thin and often subject to contentions over turf. One typical contention occurs between the ORTA and the Public Affairs Office. Given the outreach that technology transfer entails, these two functions might be assumed to be allies, but they are not necessarily so. As one ORTA explained to me, "Around here the job of Public Affairs is damage control for the commander. Nothing else. So whenever I ask for some administrative help, I get invited to put it in a memo and send it to the P.A.'s boss."

The FTTA assigns responsibility for technology transfer to all scientists and engineers in the laboratories, but the word has not yet reached them all. Although the FLC and NASA regional centers regularly run training seminars in technology transfer, the trainers frequently find participants who are totally uninformed. Although the purpose of the seminars is to present tactics of technology transfer, many participants arrive at them innocent of the reason for the whole enterprise. "They have not even been briefed on the policy of their own organizations," one old timer says. "The chief hasn't even told the indians which way to march." Anecdotes abound on this subject, always off the record.

The problem of management's tunnel vision is urgent enough to have caused the FLC to address it categorically. Beverly Berger (1994) associates the difficulties of measuring the technology transfer effort with the difficulties in its management. She cites the law and points out that it is being disregarded. The FTTA requirement is clear:

> Each laboratory director shall ensure that efforts to transfer technology are considered positively in laboratory job descriptions, employee promotion policies, and in evaluation of the job performance of scientists and engineers in the laboratory (p. 28).

"Unfortunately," she observes, "many laboratory directors have yet to implement this provision of the FTTA, even minimally." They have not done so because they have not been told unequivocally that technology transfer is an integral part of their mission. Telling them is the responsibility of DDR&E and of the generals and the admirals who dispense R&D money to the laboratories. Berger (1994) proposes accountability all the way up the line, and she invokes the performance evaluation as the time-tested management tool for achieving it: "Like the scientists and engineers who report to them, the laboratory directors and their funding organizations can be evaluated against standards of performance that make technology transfer an integral part of each laboratory's mission" (p. 29). Berger emphasizes the need to accelerate cultural change.

STASIS AND THE CORPORATE CULTURE

As everyone knows, cultural change is a slow process, and corporate culture is no exception; it has inertial force. The mission of an organization and the environment in which the organization functions determine the kind of culture it will have. That culture is a set of beliefs and values, and consequent behaviors that are passed along from one generation of employees to another; it is self-perpetuating, and that is its strength. New hires learn it and then pass it along. They perceive certain informal behavior among their managers and peers that conveys the prevailing beliefs and values of the organization. They distill those beliefs and values into behavioral rules. And if they want to survive in that environment, they adhere to the rules and become part of the communication network that teaches subsequent new employees.

Terrance Deal and Allan Kennedy (1982) have devised a schema for classifying different kinds of corporate cultures. Derived from many organizations, their representation depends on the relationship between two variables: risk and feedback. The relative risk that a company undertakes in its normal course of business is rated as being either high or low. The length of time that elapses before the company gets feedback on the success of that undertaking is rated as being either fast or slow. Since each variable has two measures, there are four possible combinations:

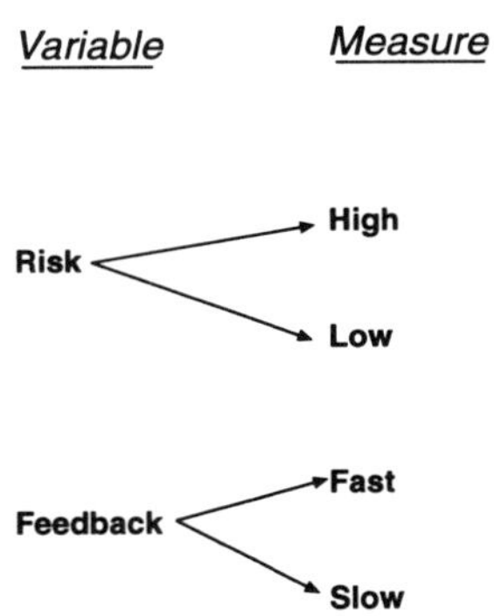

From the four possible combinations, Deal and Kennedy derive four kinds of business environments and thereby four generic cultures, which they describe. Two of the combinations can be applied to the defense R&D activities. One is a high-risk, slow-feedback environment, within which R&D normally occurs. Big stakes are riding on the business decisions, but years may pass before the outcome of the decisions is known. This combination fits the large investment and the long cycle time entailed in the development of complex weapon systems; it corresponds to the process of defense acquisition. A different combination—low risk and low feedback—corresponds to the environment of technology transfer. Deal and

Kennedy call this the *process culture.* In its extreme form it becomes a bureaucracy, wherein low risk and the lack of feedback forces employees to concentrate on method and orderly attention to detail. A bureaucracy makes its own orderly environment.

Deal and Kennedy observe that if the business environment of an organization changes, so does its culture. But the stronger the culture, the more it resists change. R&D is a high-risk business concerned with high-stake problems. Its culture centers on exhaustively analyzing the problems, factoring them into algorithmic solutions, and producing the knowledge to support those solutions. Technology transfer, on the other hand, is a low-risk operation, at least from the perspective of the information custodian; it engenders a process culture, one which deals very differently with problems. Deal and Kennedy explain that in a process culture there is a fair chance that a problem will not even be noticed. If it *is* noticed, it will probably be addressed by modifying the process. As they put it, the aim in this culture is to allow the process to manage itself toward a solution.

If the defense R&D activities are to be effective in both acquisition and technology transfer, they will need to find a way to enable the two cultures to coexist. Because new money is not available for support, they will need to grow the new culture from within. But if, as Deal and Kennedy say, the stronger culture is more resistant to change, there is a consequence here for the defense R&D activities. Since the culture of acquisition has dominated them for the past 50 years, we can assume it will resist change.

In most situations, peoples' reasons for resisting change are familiar enough. S. L. Stokes (1991) reviews them for the benefit of managers. Perhaps people do not see a reason for the change; it was either forced on them, or it happened without explanation. Or perhaps the prospect of change presents fearsome unknowns: a loss of some sort, the imposition of greater responsibility, or the possible exposure of one's lack of competence. The prospect of change, moreover, provokes the fear of a new social structure, a realignment of peers or some other invasion of one's personal values. All told, resisting change is more tempting than submitting to a new system of rewards and punishments. Given some or all of these fears, rooting a new plant from the present one may be achievable, but it will probably take a long time.

THE MARKETING MINDSET

Various formative means of assuaging fears and accelerating change are already in use in some federal laboratories. They include training seminars, reward rituals, project-oriented workshops, and exchanges of personnel between laboratories and commercial firms. These instrumental measures are effective—up to a point. But by themselves they are limited. It is likely that their effectiveness would increase, as in any teaching situation, if the learners were to know why they were engaging in these training initiatives. What seems to be needed is an understanding of the rationale for technology transfer: not only the economic or political importance of it, but also the way it works intrinsically, the way it synergizes knowledge production.

The significance of technology transfer lies in how it works. It is a mutual exchange that generates new perceptions for all participants. The task of reformulating a device or a service and adapting it to other purposes brings the supplier and the prospective user into an interactive relationship, and both parties prosper. The transfer entails a shared interpreting of old knowledge or information in order to produce new knowledge. The interpreting engages both parties in a dialogue that yields reciprocal discoveries. In this regard, I spoke earlier about the efficacy of dialogue in the R&D procurement process. When given the opportunity, dialogue accomplishes the same for technology transfer as it does for procurement.

As a dialogue proceeds, both parties reciprocate in mutual reinterpretation of the old technology, and both parties get the payoff. But this idea seems to have escaped the management of about half of the defense R&D activities. The reciprocity that occurs in the transfer process actually serves the acquisition of knowledge which is the primary mission of the defense R&D activity. It does so because it necessarily accesses a different point of view. The reciprocity also serves one of the main purposes of defense conversion by providing defense R&D activities with new perspectives of their own work. There is a simple analogy to this situation. If two parties are mutually concerned over converting a refrigerator into a heat pump, both parties will learn more about each system.

Loren Schmid (1993) addressed this concept, and the misunderstanding of it, when he spoke before the Research and Technology Subcommittee of the House Committee on Armed Services:

> Technology transfer often is discussed as if it were a one-way process from federal laboratories to U.S. industry, with the benefits flowing only to industry. In fact, technology transfer is an interactive process. The mutual stimulation that occurs with the exchange of ideas, the synergisms that spring up—these benefits accrue to both parties (p. 1).

Berger (1994) has made a different but comparable claim for this kind of interaction: "With respect to federal laboratories," she says, "defense conversion *is* technology transfer" (p. 59).

Once the rationale for technology transfer is understood, it is easier to see what is needed to make these interactions happen. What is needed is some kind of outreach, but not just any kind. Outreach is an attractive concept, but it is another one of those abstractions that conveys different meanings. Unless it is clear what kind of outreach is intended, the term can be misleading. In government circles, outreach generally means distribution of some kind; it means the dissemination of information or the dispersion of resources. But the outreach needed for technology transfer comes closer to empathy. This kind of outreach is focused on someone beyond one's self; it entails the perception of someone else's point of view. This kind of outreach is more acute than merely an awareness; it is a discernment of what someone else needs, and a new understanding based on that discernment.

What is needed for technology transfer is this kind of focused outreach, the kind that inheres in the marketing mindset. The discernment of a customer's needs is the basis of all marketing initiatives, and it becomes a means of focusing one's own capabilities. This mindset is readily at hand, of course. Marketing in the commercial sector of the economy infused the defense R&D community early in the Cold War. It has commonly characterized transactions between customer and vendor in the defense acquisition process ever since. But, like the purloined letter in Edgar Allan Poe's story, it is so obvious as to be overlooked.

Perhaps the reason for the oversight, as in Poe's mystery story, is the conventional way in which people categorize; things more or less alike tend to be associated with one another. Marketing and selling, for example, are usually lumped together in common usage; they often seem to be interchangeable. But marketing and selling, although associated, are not the same. To market a product is to provide an organizational context for selling it. The focusing kind of outreach peculiar to marketing is essential to the process of exchange. Buying and selling goods is only one version of the exchange process; technology transfer is another. But the process itself is generic.

For 40 years or more, marketing has been the means that commercial brokers use in the business of technology transfer, both nationally and internationally. Commercial brokering consists of acquiring the management of an exclusive license for a technology, finding capital, and searching for commercial opportunities by means of dialogue. This is the paradigm. Although there are many versions of it, the essential activity is aggressive networking. The various marketing initiatives all entail finding people with different kinds of resources, bringing them together, and forming a partnership with a common purpose.

The obstacles to partnering come from both parties. Analyzing the transfer process for the benefit of commercial users, Robert Spinharney (1994) formulates the need for effective transfer as a problem with a solution. The problem is the *ad hoc* manner in which knowledge is transferred from the laboratory to the market: "No universal, well-defined process," he says, "just aggressive individuals and companies discovering different ways to get and then use this plethora of knowledge and technology for their benefit" (p. 102).

Spinharney sees the problem in terms of the corporate knowledge user. He points out what information specialists have long since known. Engineers, scientists, and technical professionals tend to work in corporate isolation. Studies of their information-gathering habits indicate that, like most other people, they tend to rely on familiar sources of information. They rely on conversations with colleagues, correspondence, and a few specialized journals. Engineers tend to focus inward, as opposed to reaching out, and so do the organizations that employ them. "American companies have not had a tradition of going outside for knowledge. For many, therefore, technical outreach is counter cultural" (Spinharney, 1994, p. 102). The present culture was acceptable when technologies lasted longer, markets were more stable, and money was more important than information. But all that has changed. The urgency of the problem is that we can no longer afford the habit.

To help solve the problem there are already substantial resources for accelerating technology transfer. They include library and information networks, software products, abstracting and indexing services, and computerized searching aids. They also include information-broker services, data collection and analysis facilities, and document delivery services. These instrumental resources can minimize the difficulties of technology transfer, but only if they meet certain quality standards. Spinharney tells potential information users what they should expect from the resource services that deliver information to them. Those services must be complete, capable of responding to virtually any technical question; they must be timely and reliable. They must accommodate the user's work habits and preferences, and they must deliver the product to the user's desk.

These specifications constitute an important value added. "All users must be trained and made to feel comfortable with the service. The delivery system must be psychologically sensitive. Training must be ongoing until the service becomes routine" (Spinharney, 1994, p. 102). Thereafter, the user will become more participative and demanding of the resource service, and the transfer process will prosper. What is implicit here, and what needs to be emphasized, is that meeting these specifications is the responsibility of the information supplier.

The problem Spinharney poses is the discrepancy between outmoded work habits and the necessary speed and reliability of technology transfer that the present market demands. The old work habits of the clients ought probably to be changed. Meanwhile, reducing the discrepancy between their habits and their needs becomes the responsibility of the various information delivery services. To meet the responsibility, they must respond to the clients' questions completely, quickly, and with psychological sensitivity. It is evident that such performance qualities are all values added to the information itself.

BROKERING THE DEAL

In the defense R&D activities, effective education for technology transfer involves more than merely encouraging people to change their behavior. They might better be shown how what they already know as efficient customers can readily serve them in their new role as suppliers. They already have the capability of interpreting information, which is the essential value added. In order to convey this value to the client, the information specialist needs to interpret the client's queries with a marketing mindset.

Incorporating all this added value into the dispensing of information is a tall order, yet it is being accomplished. The achievers are a breed of professional information specialists who have acquired the marketing mindset. Starting typically from an established disciplinary base—engineering, the sciences, law, medicine, accounting, or whatever—they have extended their experiences to the process of technology transfer. They are knowledge brokers.

These are the people who manage and operate the regional centers of the FLC and NASA and also the NTTC, or they represent institutions that interact with these national networks. They belong to professional associations that represent the public, private, and educational sectors of the technology transfer enterprise. Among these associations are the Licensing Executives Society, the Association of University Technology Managers, the National Business Incubation Association, the Association of Federal Technology Transfer Executives, and the Technology Transfer Society. Some of these knowledge brokers work at technology transfer for a living; others volunteer their services. By and large, they are highly skilled people.

The emergence of an R&D broker community was natural under the circumstances. The relationship between the information custodian and the knowledge user has been marked sometimes by a symmetry of paralysis. Despite the potential benefits of their interaction, each is committed to objectives that might appear to be incompatible with technology transfer. Either party, therefore, may be tempted to resist the kinds of institutional changes that enable such transfer to happen. In this situation, the marketing mindset needed for technology transfer is notably absent. Both parties can, and do, market themselves in other ways, notably in the pursuit of capital. But curiously, with few exceptions, they have not employed this capability in technology transfer. Under these circumstances, brokering is inevitable.

The emerging community of transfer brokers has the necessary capabilities to systematize the conversion of information to knowledge. With acquired competence in information science and marketing, they can find out what resources are available, and they can access them. They also have the capability for market research. Already aware of the cultures in different market segments, they know that the accommodation of needs to resources entails the translating of cultural conventions. To use Machlup's terms, they combine many different kinds of knowing-what and knowing-how. In sum, they know what is entailed in the conversion of information to knowledge, and they know how to make the conversion happen.

Federal technology transfer has evolved into a three party system, consisting of the custodian, the user, and the broker. It has evolved to this because neither of the first two parties has the capability of making transfer happen on the scale that is needed. The notion of a three-party system has a useful analogy in the health care industry, wherein the insurance company adjudicates the transactions between physician and client. Like health insurers, moreover, knowledge brokers are engaged in improving the quality of the product; reducing risks is better for business. So the brokers are committed to the indoctrination and training of the other two parties.

Although all of the government-sponsored organizations that support technology transfer engage in some form of training, the most comprehensive commitment is undertaken by FLC. Its regional organizations oversee technology demonstrations. Headquarters publishes a broad range of reference texts and handbooks, as well as a national newsletter, and it maintains a speaker's bureau. FLC conducts training seminars at both the regional and national levels. It also regularly advises congressional committees on how technology transfer bears on their particular concerns. Virtually all of these activities are conducted by volunteers who are also employed in the federal laboratories. There is even an FLC alumni organization of volunteers who have otherwise retired.

One sure measure of learned behavior is the ability to generalize, and the knowledge brokers attempt to teach to this objective. They do more than facilitate; they engage the problem at hand in such a way as to show prospective knowledge users how to achieve a broader perspective. Each transfer is unique, of course, and there is no one universal method to achieve it. But experiences have shown that certain measures and techniques are better than others, and they can be tried in combinations that fit the local case. Through experience, brokers have learned how to locate a given situation within a larger context. From the many examples available in brokers' training programs, perhaps one will suffice to show how the local and the universal are combined.

Figure 13.1 represents a generalized map of the commercialization process. It has been used and tested by the University of Houston Clear Lake in the affiliate network of the Mid-Continent Technology Transfer Center. The matrix factors the commercialization process progressively into three major phases, six stages, and 18 significant steps. It enables the information custodian and the client to locate each technology transfer within a complex of tasks that commercializing a product normally entails. The parties can then determine what next needs to be done. The matrix can be read either down or across. *Reading down* reveals the six stages in the three major phases; *reading across* reveals successive steps in each of the six stages of the process.

Not all of these operations need engage every client. Perhaps some of them have already been completed; perhaps others do not fit the case. But some combination will be needed in each instance. The map is a kind of stop-action device, as in photography, that provides access to a process from any point within it. Even better than photography, it simultaneously reveals the whole process. With a broker to gloss the matrix by asking pertinent questions, the construct provides an agenda for use in the present instance. It also provides any user with a universe from which to choose the appropriate agenda for some future transfer. The construct is a fair example of the social technology that transfer brokers have developed, one that facilitates and teaches at the same time.

THE GENRES OF TECHNOLOGY TRANSFER

Federal technology transfer has moved swiftly since its inception. The idea of recycling technologies among federal laboratories began in 1970 with the Domestic Action Program undertaken by the Navy's R&D activity at China Lake. Now it is the broadest component, the one with the farthest reach, in the enterprise of defense conversion. During these nearly 30 years, the

THE COMMERCIALIZATION MODEL

	TECHNICAL	MARKET	BUSINESS
Concept Phase			
INVESTIGATION	TECHNICAL CONCEPT ANALYSIS	MARKET NEEDS ASSESSMENT	VENTURE ASSESSMENT
Development Phase			
FEASIBILITY	TECHNICAL FEASIBLITY	MARKET STUDY	ECONOMIC FEASIBLITY
PLANNING	ENGINEERING PROTOTYPE	STRATEGIC MARKETING PLAN	STRATEGIC BUSINESS PLAN
INTRODUCTION	PRE-PRODUCTION PROTOTYPE	MARKET VALIDATION	BUSINESS START-UP
Commercial Phase			
FULL SCALE PRODUCTION	PRODUCTION	SALES AND DISTRIBUTION	BUSINESS GROWTH
MATURITY	PRODUCTION SUPPORT	MARKET DIVERSIFICATION	BUSINESS MATURITY

FIG. 13.1. The innovation process: A model for commercialization.

concept of technology transfer has become more precisely defined. It is understood to be the dissemination of knowledge developed within one organization for one purpose to another organization for another purpose. The concept is clear.

But its implementation has encountered formidable detours within a corporate culture that reinforces the acquisition of knowledge rather than its dissemination. Most defense R&D activities were unprepared for the change from consuming knowledge to supplying it. With the lack of committed outreach among information custodians at the operational level, a new population of facilitators has materialized. What began as a niche service has become indispensable to the production of new knowledge derived from defense technologies.

Genre analysis provides a language for describing this evolution. It posits, for example, a discourse community. The knowledge brokers constitute a coherent community with a shared communicative purpose. They have built networks of information sources and resources, and they have developed ways of helping custodians and clients to engage in dialogue about needs and solutions. The brokers are slowly initiating these parties into the possibilities of a broad technology transfer community.

Genre analysis also posits forms of discourse related to shared communicative purpose. Knowledge brokers are initiating custodians and clients into functional kinds of discourse: forms of query for clients to use, forms of outreach statements for custodians, and forms of mapping the transfer process. With one notable exception, these ways of communicating are not yet genres. The exception is the CRADA; as an instrument of transfer it is now widely used, and as a document it has become highly formalized. Other forms of outreach have not yet been sufficiently generalized for information custodians to recognize them or to use them.

It may be that genre analysis will provide a way of tracking the continuing evolution of technology transfer, and possibly even forecasting it. As the marketing mindset of knowledge brokering becomes more widely understood, more information custodians may see its implications for themselves. They may perceive its compatibility with the kind of partnering they already engage in when they are acquiring knowledge from vendors. And they may come to realize the mutual payoff yielded up by the process of coproducing knowledge.

If such perceptions do occur over time, genre analysis should be able to identify the forms of discourse that the clients and custodians exchange. By relating the rationale of such records to their architectures, as we have seen in the defense acquisition process, genre analysis might be expected to produce new knowledge. It could reveal patterns, the better to initiate new custodians and clients into the dynamics of technology transfer. Or it could be a means of assessing the process and its outcomes.

As for forecasting the process at the present time, we already have some patterns to go by. The techniques and the organization of resources that knowledge brokers now use offer some reliable signals of how technology transfer will function in the near future. Small groups of information specialists with a penchant for marketing will be linked by networks to one another and to a multitude of subject specialists. Information will flow both ways between the supplier and the user. And judging by the rapid changes in the state-of-the-art in this new social technology, the knowledge output will be produced much faster than it now is.

There will be detours, of course. As at present, the brokers will operate in a jungle of unanticipated obstacles; the fit between needs and resources will not always occur. Although broker communities will doubtless devise new ways to guide and measure the flows of directed information, technology transfer is not destined to become an exact science. For that matter, neither is the forecasting of behavior in large social systems. Meantime, perhaps the most reliable advice comes from the futurist, Joseph Coates (1994): "We all have to believe we don't know what we're doing," Coates says. "Then every single encounter will be a learning experience" (p. 5). This may not be the last word, but it is durable.

Appendix: Defense Critical Technologies

As described sequentially in the *Department of Defense Critical Technology Plan* (1991), the following 15 technologies are seen as having dual-use applications. They are listed here with my annotations based on the extended descriptions in the document.

Seven other critical technologies intended to serve military applications exclusively are cited at the end of the list.

1. *Semiconductor Materials and Microelectronic Circuits* involve phenomenal reductions in the size of complex electronic circuits within computing devices in order to achieve lower power demand and cost, as well as higher reliability and speed.

2. *Software Engineering* focuses function and flexibility in most large-scale information-processing systems.

This technology involves all phases of the software life cycle—from design, through development, to use and reuse—that are needed to link different kinds of computing devices in realtime situations.

3. *High Performance Computing* signifies higher speed at lower cost in large-scale parallel computing systems.

The capability of a trillion operations per second entails the development of advanced, fault-tolerant networks of computers, enabling the assembly of encyclopedic libraries of problem domains.

4. *Machine Intelligence and Robotics* incorporates into mechanical devices certain elements of intelligent behavior, including image recognition,

knowledge representation and reasoning, learning, decision making, and planning.

This composite technology increasingly depends on advances in software design, parallel computing, composite materials, and sensor technology.

5. *Simulation and Modeling* visually represents some entity or set of entities in order to clarify their collective behavior.

It is derived from other technologies, including computer architecture, software, networking, and visualization. It is also instrumental in its own right for advancing other technologies in less time at lower cost.

6. *Photonics* uses directed light to represent, manipulate, and transmit information.

Advantaged by the speed of light, the use of laser devices and fibre-optics in integrated optical systems will someday provide order-of-magnitude improvements over the capability of electronic circuits.

7. *Passive Sensors* monitor conditions in an environment without emitting any signals. They can be located in space, under water, or within structures to monitor changes in critical variables.

Developments in this technology consist mainly of the suppression of signals that sensors emit, such as acoustic, thermal, seismic energy, or radiation.

8. Signal and Image Processing extracts relevant information from sensors. The technology entails filtering, segmenting, and classifying the information and making it accessible for automatic decision making.

New directions include the development of algorithms to reduce computational intensity and also artificial neural networks to perform pattern recognition in various applications, including robotics.

9. *Data Fusion* integrates disparate data from diverse sources of information. In so doing, it deals with the management of inherent uncertainties in those data.

The technology includes database design, search and retrieval mechanisms, and multi-media database display, all of which are instrumental to the assessment of ambiguity. For this main purpose, the development of reasoning systems is essential.

10. *Computational Fluid Dynamics* entails the calculation of fluid flow around moving bodies and the effects of its turbulence on their behavior.

The normal complexity of nonlinear differential equations with independent variables in three spatial dimensions and time is further compounded by other variables for high speeds at high altitudes. For the purpose of vehicle design in these conditions, the major thrust of this technology is the development of architecture and algorithms for massively parallel computing.

11. *Air-Breathing Propulsion*, a feature of many engine types, affects the performance of a wide range of vehicles, especially aerospace vehicles.

Two kinds of design, aerothermodynamic and structural, represent the major thrusts of this technology, along with the development of endothermic fuels and high-temperature materials for hypersonic flight. As in the technology of fluid dynamics, design depends on computational methods of analyzing interactive nonlinear physical phenomena.

12. *Composite Materials* have selected properties superior to those of their component parts.

Combining polymer, metal, carbon, or ceramic with other materials to meet severe thermal, structural, and mechanical requirements is considered to be the enabling technology for deep submergence vessels, certain spacecraft, and long-range missiles. Composites have widespread applications in many other structures as well.

13. *Superconductivity* is the property in certain materials of drastically reducing electrical resistance and consequent power loss in electric, electronic, and electromagnetic devices.

The technology based on this property focuses on both low-temperature and high-temperature superconductive materials, and it broadly extends from theoretical understanding of the phenomenon to test models and operational systems.

14. *Biotechnology* exploits genetic mechanisms that control living organisms and puts them to a variety of uses.

The technology includes the engineering of cells that act as factories for synthesizing new materials at low cost, the synthesizing of enzymes to catalyze the degradation of toxic chemicals and hazardous wastes, and the fabricating of exquisite sensors that can detect a single molecule or a single photon.

15. *Flexible Manufacturing* is a set of processes that enables efficient production to occur despite sudden changes in production volume or in the product itself.

The main elements of this technology are precise product definition and a work force trained to respond to change. The tools that implement it are computer-assisted design and manufacturing software, sensors, database management, and intelligent software interfaces.

The seven critical defense technologies intended to serve military applications exclusively are:

Sensitive radars,
Signature control,
Weapon system environment,
Pulsed power,
Hypervelocity projectiles and propulsion, and
High energy density and materials.

References

Abelson, P. H. (1991, February 8). Federal impediments to scientific research. *Science, 251*, (4994), 605.

Aeronautical Systems Division, AFSC. (1990). *Long Range Acquisition Estimates, FY91 and FY92*. Wright-Patterson Air Force Base, OH: Author.

Allen, T. J. (1977/1986). *Managing the flow of technology*. Cambridge, MA: MIT Press.

Allison, D. K. (1985). U.S. Navy research and development since World War II. In M. R. Smith (Ed.), *Military enterprise and technological change*. Cambridge, MA: MIT Press.

American Forces Information Service. (1994). *Defense almanac, Issue 5*. Washington, DC: U.S. Government Printing Office.

American Society of Mechanical Engineers. (1986). *Goals and priorities for research in engineering design: A report to the design research community*. New York: Author.

ARPA. (1995). *The technology reinvestment project: Dual-use innovation for a stronger defense*. Arlington, VA: Author.

Augustine, N. R. (1983). *Augustine's laws* (revised and enlarged). New York: American Institute of Aeronautics and Astronautics.

Barke, R. P. (1991, October). *Politics and the federal patronage of science*. Paper presented at the annual meeting of the Association for Public Policy Analysis and Management, Washington, DC.

Barrett, R. (1994, Spring). Going for brokers. *Technology Transfer Business*, 35–38.

Baucom, D. R. (1992). *The origins of SDI,1944–1983*. Appleton, WI: Lawrence University Press of Kansas.

Bazerman, C. (1988). *Shaping written knowledge: The genre and activity of the experimental article in science*. Madison: University of Wisconsin Press.

Bazerman, C. (1993). Money talks. In W. Henderson, T. Dudley-Evans, & R. Backhouse (Eds.), *Economics and language* (pp. 153–199). London: Routledge.

Bell, D. (1976). *The coming of the post industrial society*. New York: Basic Books.

Berger, B. J. (1994, August). *Technology transfer in a time of transition: A guide to defense conversion*. Washington, DC: Federal Laboratory Consortium.

Bjork, R. S. (1992). *The strategic defense initiative: Symbolic containment of the nuclear threat*. Albany: State University of New York Press.

Bolos, J. T. (1991, February). Innovative procurement concepts—The broad agency announcement. *Contract Management*, 4–8.

Boone, L. E., & Kurtz, D. L. (1980). *Contemporary marketing*. Hinsdale, IL: Dryden Press.

Branscomb, L. (1995). *Confessions of a technophile*. Woodbury, NY: American Institute of Physics.

Bush, V. (1945/1960, July). *The endless frontier: A report to the president on a program for postwar scientific research.* Reprinted, National Science Foundation, *NSF 60–40,* July 1960.

Business Week. (1992, August 3). "Smart selling: How companies are winning over today's tougher customer." *Business Week, 3277,* 47–52.

Cairncross, A. (1976). The market and the state. In T. Wilson & A. S. Skinner (Eds.), *The market and the state: Essays in honour of Adam Smith* (pp. 113–135). Oxford, England: Clarendon.

Carter, M. (1988, September). Problem solving reconsidered. College English 50, (5), 551–565.

Center for Defense Information. (1996, February). Why we overfeed the sacred cow. *The Defense Monitor, 25,* (2), 1–7.

Chace, J., & Carr, C. (1988). *America invulnerable: The quest for absolute security from 1812 to star wars.* New York: Summit.

Coates, J. (1994, April). The year 2025: A futurist looks back on U.S. technology, society, and economy. *T'Squared, 19,* (4), 1f.

Committee on Definitions. (1960). *Marketing definitions: A glossary of marketing terms.* Chicago: American Marketing Association.

Competitiveness Policy Council. (1994, May). *Promoting long-term prosperity: Third report to the president and congress.* Washington, DC: U.S. Government Printing Office.

Davey, M. E. (1994, December 12). DOD Support for university based research. *CRS Report for Congress.* Washington DC: Congressional Research Service, Library of Congress.

Deal, T. E., & Kennedy, A. A. (1982). *Corporate cultures: The rites and rituals of corporate life.* Reading, MA: Addison-Wesley.

Defense Conversion Commission. (1992, December 31). *Adjusting to the drawdown.* Washington, DC: Office of the Secretary of Defense.

Department of the Army, Research, Development, and Acquisition. (n.d-a). *Army technology base master plan overview.* Washington, DC: Author.

Department of the Army, Research, Development, and Acquisition. (n.d-b). *50 years of accomplishments in army research and development.* Washington, DC: Author.

Department of Defense. (1995). *Dual-use technology.* Washington, DC: Author.

DeLauer, R. (1986, July). "An Interview: Richard DeLauer on defense." *Technology Review,* 89, (5), 58ff.

Denman, G. L. (1992, March 19). *Statement before the Subcommittee on Defense, House Appropriations Committee.* Washington, DC: U.S. House of Representatives.

de Solla Price, D. J. (1965). Is technology historically independent of science? A study of statistical historiography. *Technology and Culture,* 6, (4), 553–568.

Deutsch, J. (1993, October 4). Making the Pentagon buy smarter, Deutsch ponders the changes ahead. *Defense Week, 14,* (39), 8ff.

Director, Defense Research and Engineering, Department of Defense. (1969). *Project HINDSIGHT.* Springfield, VA: National Technical Information Service.

Director, Defense Research and Engineering, Department of Defense. (1979). *DODINST 7720.16, Part V, Budget and Accounting Classifications, Attachment* C, 516f. Washington, DC: Author.

Director, Defense Research and Engineering, Department of Defense. (1991, July 22). *The Department of Defense critical technologies plan* (AD-234 900). Springfield, VA: National Technical Information Service.

Director, Defense Research and Engineering, Department of Defense. (1994a, September). *Defense science and technology strategy.* Washington, DC: Author.

Director, Defense Research and Engineering, Department of Defense. (1994b, September). *Defense technology plan.* Washington, DC: Author.

Directorate of Information Operations and Reports. (1991a). *Department of Defense 100 companies receiving the largest dollar volume of prime contract awards* (PO1). Washington, DC: U.S. Government Printing Office.

Directorate of Information Operations and Reports. (1991b). *500 contractors receiving the largest dollar volume of prime contract awards for RDT&E* (PO2). Washington, DC: U.S. Government Printing Office.

Directorate of Information Operations and Reports. (1991c). *Department of Defense prime contract awards* (PO3). Washington, DC: U.S. Government Printing Office.

Doig, L. L. (1994). *FLC history: Establishing a national presence.* Washington, DC: Federal Laboratory Consortium.

Drucker, P. F. (1954). *The practice of management.* New York: Harper & Row.

Drucker, P. F. (1989). *The new realities.* New York: Harper & Row.

Drucker, P. F. (1993). *The post capitalist society.* New York: Harper Collins.

Eliot, T. S. (1963). *Collected poems 1909–1962.* New York: Harcourt Brace.

Faulkner, W. (1994, Autumn). Conceptualizing knowledge used in innovation: A second look at the science-technology distinction and industrial inovation. *Science, Technology, & Human Values,19,* (4), 425–458.

Federal Laboratory Consortium. (1994, August). *Winners in technology transfer.* Washington, DC: Author.

Forster, E. M. (1927/1954). *Aspects of the novel.* New York: Harcourt Brace & World.

Fowler, A. (1982). *Kinds of literature.* Oxford, England: Oxford University Press.

Galbraith, J. K. (1967). *The new industrial state.* Boston: Houghton Mifflin.

Gansler, J. S. (1989). *Affording defense.* Cambridge, MA: MIT Press.

General Electric Corporation. (1952). *1952 Annual report.* New York: Author.

Georgia Tech Research Institute. (n. d.) [capability statements for laboratories named:] Aerospace and transportation; Arlington research; Electronic systems; Electro-optics, environment and materials; Hunstville research; Information technology and telecommunications; Sensors and electromagnetic applications; Signatures technology. Atlanta: Author.

Gerstenfeld, A. (1977). *Innovation: A study of technological policy.* Cambridge, MA: MIT Press.

Greenberg, D. (1994, September 15). On R&D items, big and small, congress gushes with advice. *Science and Government Report, XXIV,* (14), 1f.

Grene, M. (1985). Perception, interpretation, and the sciences. In D. Depew & B. Weber (Eds.), *Evolution at a crossroads* (pp. 1–20). Cambridge, MA: MIT Press.

Gugliotta, G. (1993, December 13–19). Railroading that pork through congress. *The Washington Post National Weekly Edition, 24.* Washington, DC: The Washington Post.

Hayes, K. C. (Ed.). (1992). *Tapping federal technology: Inventions, expertise, and facilities.* Federal Laboratory Consortium Handbook Series, No. 3. Washington DC: Federal Laboratory Consortium.

Hibbard, C. G. (1995, Fall). Training the Atomic Bomb Group. *Air Power History, 42*(3), 24–33.

Hoey, M. (1983). *On the surface of discourse.* London: Allen & Unwin.

Human Systems Center, AFMC. (1993). *Human Systems Center products and progress.* Brooks AFB, TX: Author.

Jenkins, K., Jr. (1993, December 13–19). Jurists' prudence goes on trial. *The Washington Post National Weekly Edition, 31.* Washington, DC: The Washington Post.

Joint Services Electronics Program (JSEP). (1985). *40th Anniversary of the Joint Services Electronics Program.* Arlington, VA: Anser.

Jordan, M. P. (1984). *Rhetoric of everyday English texts.* London: Allen & Unwin.

Kieth, R. J. (1960). The marketing revolution. *Journal of Marketing, 24*, 35–38.

King, A. (1993, October). Is cost accounting a threat to our national security? *Management Accounting*, 47–51.

Kinnear, T. C., & Bernhardt, K. L. (1983). *Principles of marketing.* Glenview, IL: Scott, Foresman.

Kline, S. J. (1986, January). What is technology? *S-STS Reporter, 2*, (1).

Knorr-Cetina, K. D. (1981). *The manufacture of knowledge: An essay on the constructivist and contextual nature of science.* New York: Pergamon.

Kotler, P. (1980/1983). *Principles of marketing.* Englewood Cliffs, NJ: Prentice-Hall.

Kuhn, T. (1962/1970). *The structure of scientific revolutions.* Chicago: University of Chicago Press.

Latour, B. (1987). *Science in action.* Cambridge, MA: Harvard University Press.

Lakoff, S., & York, H. F. (1989). *A shield in space? Technology, politics, and the strategic defense initiative.* Berkeley: University of California Press.

Lawrence, R. M. (1987). *Strategic Defense Initiative: Bibliography and research guide.* Boulder, CO: Westview Press.

Lee, J. W. (1993, July-August). Changing role of government laboratories: Improving U.S. industrial competitiveness. *Program Manager*, 48–51.

LeSueur, S. C. (1944, August 15–21). DoD, industry face hurdles in elimination of milspecs. *Defense News*, 22f.

Lippmann, W. (1943). *U.S. foreign policy: Shield of the republic.* Boston: Little, Brown.

Longino, H. E. (1990). *Science as social knowledge.* Princeton, NJ: Princeton University Press.

Machlup, F. (1980). *Knowledge and knowledge production.* Princeton, NJ: Princeton University Press.

Maraniss, D., & Weiskopf, M. (1995, September 24). In test of strength stealth survives. *The Washington Post*, 1f.

Marschak, T., Glennan, T. K., Jr., & Summers, R. (1967). *Strategy for R&D: Studies in the microeconomics of development*, Rand Corporation Research Study: Number 8 in the series, Econometrics and Operations Research. New York: Springer-Verlag.

Massachusetts Institute of Technology, Center for Policy Alternatives. (1978). *Government involvement in the innovation process* (contractor's report to the Office of Technology Assessment). Washington, DC: U.S. Government Printing Office.

Meeks, R. L. (1994). *Federal R&D funding by budget function: Fiscal years 1993–1995*, NSF 94–319. Arlington, VA: National Science Foundation.

Mintz, J. (1993, November 15–21). The new guard in defense. *The Washington Post National Weekly Edition*, 32.

Misa, T. J. (1985). Military needs, commercial realities, and the development of the transistor, 1948–58. In M. R. Smith (Ed.), *Military enterprise and technological change.* Cambridge, MA: MIT Press.

Myers, G. (1990). *Writing biology.* Madison: University of Wisconsin Press.

Naisbitt, J. (1982). *Megatrends.* New York: Warner Books.

National Economic Council, National Security Council, & Office of Scientific and Technology Policy. (1995, February). *Second to none: Preserving america's advantage through dual-use technology.* Washington, DC: Author.

National Science Board. (1993). *Science and engineering indicators: 1993*, NSB 93-1. Washington, DC: U.S. Government Printing Office.

Naval Air Propulsion Center. (1991, July). *Guide to naval aircraft gas turbine propulsion technology development programs for fiscal year 1992*, NAPCAPT-92. Trenton, NJ: Author.

Naval Undersea Warfare Center, Newport Division. (1994). *Science and technology 1994.* Newport, RI: Naval Undersea Warfare Center.

Newell, A., & Simon, H. A. (1972). *Human problem solving*. Englewood Cliffs, NJ: Prentice-Hall.

Nunn, J. (1994). *Assessing the potential for civil-military integration: Technologies, processes, and practices*. Washington, DC: U.S. Government Printing Office.

Office of the Secretary of Defense. (1979). *DODINST 7720.16. Part V, Budget and Accounting Classifications, Attachment* C (516-1f.). Washington, DC: Author.

Office of the Secretary of Defense. (n.d.). *Department of Defense in-house RDT&E activities: Management analysis report for fiscal year 1990*. Washington, DC: Author.

Office of the Secretary of Defense. (1994). *Department of Defense in-house RDT&E activities: Management analysis report for fiscal year 1992*. Washington, DC: Author.

Office of Technology Assessment. (1988, March). *The defense technology base: Introduction and overview—A special report*, OTA-ISC-374. Washington DC: U.S. Government Printing Office.

Office of Technology Assessment. (1989, April). *Holding the edge: Maintaining the defense technology base*, OTA-ISC-420. Washington DC: U.S. Government Printing Office.

Office of Technology Assessment. (1993, May). *Defense conversion: Redirecting R&D*, OTA-ITE-552. Washington, DC: U.S. Government Printing Office.

Office of Technology Assessment. (1995). *A history of the Department of Defense federally funded research and development centers*. Washington, DC: Author.

Ogden Air Logistics Center. (n.d.). *Integrating tomorrow's technology today*. Hill AFB, UT: Author.

Otteson, S. F., Panschar, W. G., & Patterson, J. M. (1964). *Marketing: The firm's viewpoint*. New York: Macmillan.

Pride, W. M., & Farrell, O. C. (1977/1980). *Marketing: Basic concepts and decisions*. Boston: Houghton Mifflin.

Reagan, R. (1983a, March 8). Remarks at the annual convention of the national association of evangelicals in Orlando, FL. *Public Papers of the Presidents: Ronald Reagan, 1983, Book 1*, 359–364.

Reagan, R. (1983b, March 23). Address to the nation on defense and national security. *Public Papers of the Presidents: Ronald Reagan, 1983, Book 1*, 437–443.

Reis, V. H. (1991a, April 23). *Statement before the Subcommittee on Research and Development, Armed Services Committee, U.S. House of Representatives*. Washington, DC: U.S. House of Representatives.

Reis, V. H. (1991b, November–December). Foundations for the future. *Defense* , *91*, 21–26.

Reingold, N. (1991). Science and government in the united states since 1945. In *Critical problems and research frontiers* (working papers for a conference of the History of Science Society and Society for the History of Technology, Madison, WI, October 30–November 3, 1991).

Ricks, T. E. (1993, May 20). Shot down: With the cold war over, the military-industrial complex is resolving. New York: *The Wall Street Journal* 1f.

Ritenburg, P. J. (1988, Summer). Task order contracts: Popular but are they legal? *NCMA Journal*, 35–45.

Roberts, E. B. (1965). *Marketing and engineering strategies for winning R&D contracts*. Paper presented at the 4th annual Management Conference on Marketing in the Defense Industries, Boston, May 27, 1965.

Roberts, E. B. (1967). *The measurement and improvement of R&D marketing effectiveness*. Paper presented at the 5th annual Conference on Marketing in the Defense Industries, Boston, May 19, 1966.

Roessner, D., & Wise, A. (1993). *Patterns of industry interaction with federal laboratories*. Atlanta: Georgia Institute of Technology, School of Public Policy.

Ross, I. (1993, January 17). R& D: Right money, wrong topics. *New York Times*, Section 3, 13.

Rowberg, R. E. (1995, December 15). *Federal research and development funding: A concise history*, 95-1209 SPR. Washington, DC: Congressional Research Service, Library of Congress.

Sapolsky, H. M. (1972). *The Polaris system development: Bureaucratic and programmatic success in government*. Cambridge, MA: Harvard University Press.

Schlaifer, R. (1950). *Development of aircraft engines*. Boston: Harvard University, Graduate School of Business Administration.

Schmid, L. C. (1993, April 21). *Verbal statement before the Research and Technology Subcommittee of the House Committee on Armed Services*. Washington, DC: Federal Laboratory Consortium.

Schmoll, J. (1993). *Introduction to defense acquisition management*. Fort Belvoir, VA: Defense Systems Management College Press.

Shapley, D., & Roy, R. (1985). *Lost at the frontier: U.S. science and technology policy adrift*. Philadelphia: ISI Press.

Shrum, W. (1985). *Organized technology*. West Lafayette, IN: Purdue University Press.

Smith, A. (1776/1976). In E. Cannan (Ed)., *An inquiry into the nature and causes of the wealth of nations*. Chicago: University of Chicago Press.

Spinharney, R. (1994). Knowledge networking: An emerging source of professional power. In *1994 Biological Sciences Symposium Proceedings* (pp. 101–103). Atlanta: Technical Association of the Pulp and Paper Industry.

Squires, A. M. (1986). *The tender ship*. Boston: Birkhauser.

Stevanus, M. (1994). *Federal technology transfer opportunities on the internet*. Baltimore: National Agricultural Library, Technology Transfer Information Center.

Stewart, I. (1948/1980). *Organizing scientific research for war: The administrative history of the office of scientific research and development*. New York: Arno Press.

Stokes, S. L., Jr. (1991). *Controlling the future: Managing technology-driven change*. Boston: QED Technical Publishing Group.

Swales, J. (1990). *Genre analysis: English in academic and research settings*. Cambridge, England: Cambridge University Press.

Swetnam, G. F., Jr. (1994, July). A capitol experience. *IEEE Spectrum*, 40–43.

Sylos-Labini, P. (1976). "Competition: The product markets." In T. Wilson & A. S. Skinner (Eds.), *The market and the state: Essays in honour of Adam Smith* (pp. 200–232). Oxford: Clarendon Press.

Technology Innovation: Chapter 63, United States Code Annotated Title 15, Commerce and Trade: Sections 3701 to 3715 (as amended through 1990 public law and with annotations). (1994). St. Paul, MN: Produced for the West Federal Laboratory Consortium for Technology Transfer, Washington, DC.

Toner, M. (1995, April 8). Food for tomorrow's army. *The Atlanta Journal-Constitution*, E1.

Tornatzky, L. G., & Fleischer, M. (1990). *The processes of technical innovation*. Lexington, MA: Lexington Books.

Toulmin, S. (1972). *Human understanding*. Princeton, NJ: Princeton University Press.

Tushman, M. L., & Moore, W. L. (Eds). (1988). *Readings in the management of innovation*. New York: Ballinger.

Undersecretary of Defense for Acquisition and Technology. (1995, February). *Dual use technology: A defense strategy for affordable, leading-edge technology*. Washington, DC: Author.

U.S. General Accounting Office. (1991, May 16). *Strategic defense initiative program: A look at lessons learned during SDIO's first seven years* (GAO/T-NSIAD-91-33). Washington, DC: Author.

U.S. General Accounting Office. (1993, July 16). *Ballistic missile defense: Evolution and current issues* (GAO/NSIAD-93-229). Washington, DC: Author.

U.S. General Accounting Office. (1994, October). *Acquisition reform: DOD begins program to reform specifications and standards*, GAO/NSIAS-95-14. Washington, DC: Author.

Vincenti, W. G. (1990). *What engineers know and how they know it: Analytical studies from aeronautical history*. Baltimore: Johns Hopkins University Press.

Viner, J. (1928). "Adam Smith and laissez-faire." In *Adam Smith 1776–1926: Lectures to commemorate the sesquicentenniel of the publication of the wealth of nations*. Chicago: University of Chicago Press.

Walker, M. (1993). *The cold war: A history*. London: Fourth Estate. (NY: Henry Holt, 1994)

Weeks, J. (1995, Fall). Congressional proponents of star wars are working to revive the program yet again. *Nucleus: The magazine of the Union of Concerned Scientists, 17*, (3), 1–3.

Wells, W. G., Jr. (1992). *Working with congress*. Washington, DC: American Association for the Advancement of Science, and the Carnegie Commission on Science, Technology, and Government.

White House. (1994, July). *A national security strategy of engagement and enlargement*. Washington, DC: U.S. Government Printing Office.

Whitman, W. (orig. pub. 1855; 1950). "Song of Myself," stanza 51. In J. Kouwenhoven (Ed.), *Leaves of grass*. New York: Random House, Modern Library.

Wright Laboratory Air Force Base. (n.d.). *Wright laboratory heritage*. Dayton, OH: Author.

Wriston, W. B. (1992). *The twilight of sovereignty*. New York: Scribner's.

Zegveld, W., & Enzing, C. (1987). *SDI and industrial policy: Threat or opportunity?* New York: St. Martin's Press.

Author Index

Subject Index

D

E

F